ENERGY POLICIES, POLITICS AND PRICES

IONIZING RADIATION

APPLICATIONS, SOURCES AND BIOLOGICAL EFFECTS

ENERGY POLICIES, POLITICS AND PRICES

Additional books in this series can be found on Nova's website under the Series tab.

Additional e-books in this series can be found on Nova's website under the e-book tab.

ENERGY POLICIES, POLITICS AND PRICES

IONIZING RADIATION

APPLICATIONS, SOURCES AND BIOLOGICAL EFFECTS

EDUARD BELOTSERKOVSKY
AND
ZIVEN OSTALTSOV
EDITORS

Nova Science Publishers, Inc.
New York

For permission to use material from this book please contact us:
Telephone 631-231-7269; Fax 631-231-8175
Web Site: http://www.novapublishers.com

NOTICE TO THE READER

Additional color graphics may be available in the e-book version of this book.

Library of Congress Cataloging-in-Publication Data

ISBN: 978-1-62257-343-1

Published by Nova Science Publishers, Inc. † New York

CONTENTS

PREFACE

Ionizing radiation is radiation composed of particles that individually can liberate an electron from an atom or molecule, producing ions, which are atoms or molecules with a net electric charge. These tend to be especially chemically reactive, and the reactivity produces the high biological damage caused per unit of energy of ionizing radiation. This new book examines the applications, sources and biological effects of ionizing radiation with a focus on the analytical methods for the identification of irradiated foods; protein kinases as fundamental participants in the response to DNA damage from ionizing radiation; radiative defects in nanocyrstals as compared with those in bulk crystals; mechanical and optical properties of ionic crystals; mtDNA as a marker of radiation damage; antioxidant prophylaxis of radiation stress; and dosimetry characterization of a cobalt-60 y-irradiation facility for the radiation sterilization of insects.

Chapter 1 - Extensive research over the past several decades has established not only the technical feasibility of food irradiation but also the wholesomeness of irradiated food. However, lack of consumers' understanding and international consensus regarding irradiated food resulted in mandatory labeling and monitoring requirements. Efficient and reliable identification methods make it possible to check the compliance with existing regulations and facilitate international trade of irradiated foods. Identification methods are also important in enhancing consumers' confidence and protect their right of choice. An ideal identification method should provide radiation-specific, sensitive, reproducible, dose-dependent, and stable results with negligible changes during post-irradiation period and/or after other subsequent food processing treatments. Current techniques are based on physical, chemical, biological, and microbiological changes in irradiated foods. However, none of the available methods have the potential to effectively characterize all food materials in accordance to their irradiation history and usually multiple methods are applied depending on the nature of the food item and available resources.

This chapter describes the potential of various analytical methods for identifying irradiation status of food products. Special emphasis is given to the validated EN standard methods, whereas the applications of other techniques with limited potential are also discussed. The requirements, technical effectiveness and practical limitations in applying these methods for a routine food control are also addressed.

Chapter 2 - Protein kinases are fundamental participants in the response to DNA damage from ionizing radiation and other insults. The molecular roles of the PI3-like kinases ATM and ATR have been well characterized in the cascade of events that detect damaged DNA, activate cell cycle checkpoints, orchestrate and amplify mediators of the response, and ensure that damaged DNA is repaired before cells divide. Another mammalian protein kinase, NEK1 (never-in-mitosis related kinase 1), has similarly important but distinct roles in DNA damage responses. Studies *in vitro*, in cells, and in animals indicate that NEK1 functions uniquely as a sensor and mediator of the response to DNA damage. NEK1 is important for limiting cell death after DNA damage, activating S-phase and mitotic checkpoints properly, ensuring faithful chromosome segregation, and preventing specific neoplastic diseases. Data suggest that NEK1 deserves to be investigated further in exploring the mechanisms that lead to aneuploidy, aberrant cell death, and uncontrolled proliferation in human diseases such as kidney cancer, lymphomas, polycystic kidney disease, and bone diseases.

Chapter 3 - Bulk crystals and nanocrystals of lithium fluoride are irradiated by various doses of gamma rays at a temperature of 77 K. The time evolution of photoluminescence from F_2^+, F_2, F_3^+ and F_3 color centers are measured at various annealing temperatures. It has been revealed that in many cases the kinetics of the reactions, determined by mobile defects diffusion in crystals, is described by the exponential dependence. At diffusion of random walk type, on the base of such dependence distribution of the diffusion pathways travelled by the mobile defects prior to entry into reactions, has been obtained. Lifetimes of anionic vacancies υ_a and F_2^+ centers, also activation energies and coefficients of their diffusion are determined. It is found that in the bulk crystals lifetime decreases for vacancies while increases for F_2^+ centers by increasing the irradiation dose. It is also shown that, after irradiation during crystals annealing, vacancies are formed as a result of the reaction $F_2^+ + H \rightarrow \upsilon_a + Fl^-$, where Fl^- is a fluorine ion in a lattice site and H is a fluorine interstitial atom. The presence of F_1^- centers in the irradiated samples is established, and the processes, which lead to the formation of F_2, F_3^+ and F_3 centers after irradiation, are discovered. It has been found that diffusion activation energy for vacancies in nanocrystals is twice as high as compared with bulk crystals. Migration of F_2^+ centers in nanocrystals has not been detected. Presence of new type of radiative defects in nanocrystals as compared with those in bulk crystals has been experimentally revealed.

Chapter 4 - The review covers the main body of works carried out from 1975 to 2012 years. They are unified by the general idea – to study experimentally effect of ionizing radiation combined with various external fields (mechanical stress, electrical and magnetic fields etc.) upon optical-mechanical properties of LiF crystals. It was shown that elementary radiation process when combined with various external fields differs essentially from the corresponding process under the same irradiation conditions but without these fields. Particularly, the additional (apart from crystallographic) anisotropy of mechanical and optical properties, caused by formation of oriented anisotropic radiation defects (bi-vacancies, F_2^+-centers etc.) and preferred slip planes, is induced.

The idea to combine various external fields was also used when studying post-radiation phenomena in crystals irradiated in reactor (annealing with uniaxial compression, hard UV irradiation with mechanical load etc.). Results of magneto-plastic effect studies in crystals under combined action of weak magnetic field and X-raying, which demonstrate the role of spin-dependent transitions in the formation of post-radiation properties of crystals, are presented.

Chapter 5 - Most studies of radiation-induced changes in genetic material of cells are based on an analysis of nuclear DNA. Lately there has been growing interest of researchers in the state of mitochondria and their genome in tissues after genotoxic influences on the organism, in as much as the stability of mitochondrial DNA (mtDNA), carrying genes of the system of energy biogenesis, is extraordinarily significant for the processes of cell restoration. MtDNA in the functioning and organization of structure has a number of peculiarities distinct from nuclear DNA (nDNA). It is replicated independently of the cell cycle and replication of nDNA. In mitochondria there is limited functioning of various systems of DNA repair. Apart of that, mtDNA does not contain noncoding sequences, in the structure of its genes there are no introns, and it is transcribed as a unified polycistronic block. In mtDNA there arise significantly more damages than in a commensurate fragment of nDNA, both as a result of influence of reactive oxygen species (ROS) generated in mitochondria themselves and upon the action on the cells of physical and chemical genotoxic agents. All this favors formation in mtDNA of disturbances/mutations at a high frequency (see review). Therefore, some researchers use mtDNA as a marker of radiation damage by analyzing changes in the number of copies, presence of deletions and mutations in mtDNA in tissue cells, and amount of mitochondrial genome fragments in the plasma / serum.

Chapter 6 - Once atomic power engineering has become a part of our everyday lives, the special precautions should be taken to reduce harmful effects of radiation for specialists in atomic industry as well as for people in the contaminated territories after atomic accidents. To defend the people in case of chronic radiation, novel radiation protectors, which would be non-toxic and suited to long-time applications as nutrients, are required. The water-soluble antioxidants based on alkyl-substituted hydroxypyridines have been long used with success in medical practice. In the experiments with yeast cells, *S. cerevisiae*, revealed that 3-hydroxy-6-methyl-2-ethylpyridine essentially improves post-radiation recovery and raises survivability of the cells after γ-irradiation (^{60}Co, 800 Gy). Of special interest, can be some magnetic isotopes. Among three stable isotopes of magnesium, ^{24}Mg, ^{25}Mg, and ^{26}Mg with natural abundance approximately 79, 10, and 11 %, only ^{25}Mg has the nuclear spin (I = 5/2) and, hence, the nuclear magnetic field, while ^{24}Mg and ^{26}Mg have no nuclear spin (I = 0) and magnetic field. Chapter 6 reveals that the rate constant of post-radiation recovery of cells after short-wave UV irradiation was twice higher for the cells enriched with magnetic ^{25}Mg, when compared to the cells enriched with the nonmagnetic isotope. Thus, the stable magnetic isotope of magnesium, as well as the non-toxic antioxidants, hold promises for creating novel radio-protectors suitable as nutrients for use at chronic radiation. [Supported by Russian Foundation for Basic Research, grant 10-03-01203a].

Chapter 7 - Ionizing radiation (IR) occurs naturally at low doses. Mostly, when IR induces DNA damage, our cells are able to repair the error. These damages can be direct, causing ionization - ejection of electrons from molecules - of DNA atoms, or indirect when there is an interaction with water and other cellular molecules, leading to a generation of reactive oxygen species.

When the cells are not able to fix the error, a sequential multistep process generates several types of chemically stable lesions, which can lead to cell death. Therefore, IR is being used for cancer treatment, reaching 50% of all tumor types. New techniques and combined therapies with radiosensitizers and chemotherapics are being evaluated in order to improve the efficacy of IR treatment. In fact, significant progress has been made in our understanding of the basic mechanisms of radiation injury and its cellular processing in both normal and malignant cells. New discoveries offer increasing opportunities for clinical applications. Our increasing understanding of the basic mechanisms controlling the cell cycle and apoptosis provides important molecular markers for the caracterization of injury responses in different types of normal and malignant cells, and important molecular targets for future therapeutic intervention. This chapter discusses some of these issues in the context of their relevance to the clinical effects of radiation on tumor and normal cells.

A detailed understanding of the mechanisms and pathways of radiation injury and repair may lead to the design of new biologic or chemical response modifiers to improve the therapeutic ratio of radiation treatments in human cancer.

Chapter 8 - Cobalt-60 gamma irradiation Pilot Plant has been put into operation in 1999 at the National Centre of Nuclear Sciences and Technology (CNSTN), Sidi-Thabet, Tunisia. An initial characterization of this Pilot Plant was performed in order to control technical specification and to determine the overall performance of the irradiator in delivering absorbed dose for sterilization of medical devices and food irradiation. A new irradiation holder was recently installed; it was designed especially for the irradiation of pupae of the Mediterranean fruit fly. It consists of four turn plates which makes it possible to rotate the canisters holding the pupae within the radiation field. The axis of rotation is vertical and parallel to the source pencils. Prior to routine irradiation using the new irradiation holder, validation procedures are necessary to establish conditions of the irradiation within the specification. In the course of these procedures, detailed dose mapping on a vertical plane in the middle of the canister of insect pupae with bulk density of 0.446 g/cm^3 was carried out for two irradiation configurations: unturned plates and turned plates. GafChromic dosimeter calibrated against Alanine /ESR dosimetry system was used for the dose measurements. The maximum and minimum dose locations were determined and the dose uniformity ratio calculated and discussed. Detailed analyses of the isodose curves and histogram of the frequency distribution of absorbed dose were also given. Transit dose and dose rate in the reference position inside the canister were measured using Fricke dosimeters. The results of measurements of absorbed dose and dose distribution in insect pupae do not show any significant difference in the dose uniformity ratio $(D.U.R. = \frac{D_{max}}{D_{min}})$ between the two irradiation configurations. At the same time it was observed with turned plate's configuration an improvement of the homogeneity of the absorbed dose distribution in the insect pupae showed by the increasing of the pupae irradiated at the minimum dose by about 17 %.

In: Ionizing Radiation
Editors: Eduard Belotserkovsky and Ziven Ostaltsov

ISBN: 978-1-62257-343-1

Chapter 1

ANALYTICAL METHODS FOR THE IDENTIFICATION OF IRRADIATED FOODS

Kashif Akram, Jae-Jun Ahn and Joong-Ho Kwon*
Department of Food Science and Technology
Kyungpook National University, Republic of Korea

ABSTRACT

Extensive research over the past several decades has established not only the technical feasibility of food irradiation but also the wholesomeness of irradiated food. However, lack of consumers' understanding and international consensus regarding irradiated food resulted in mandatory labeling and monitoring requirements. Efficient and reliable identification methods make it possible to check the compliance with existing regulations and facilitate international trade of irradiated foods.

Identification methods are also important in enhancing consumers' confidence and protect their right of choice. An ideal identification method should provide radiation-specific, sensitive, reproducible, dose-dependent, and stable results with negligible changes during post-irradiation period and/or after other subsequent food processing treatments. Current techniques are based on physical, chemical, biological, and microbiological changes in irradiated foods. However, none of the available methods have the potential to effectively characterize all food materials in accordance to their irradiation history and usually multiple methods are applied depending on the nature of the food item and available resources.

This chapter describes the potential of various analytical methods for identifying irradiation status of food products. Special emphasis is given to the validated EN standard methods, whereas the applications of other techniques with limited potential are also discussed.

The requirements, technical effectiveness and practical limitations in applying these methods for a routine food control are also addressed.

---

* Corresponding author: Tel.: +82 53 950 5775; Fax: +82 53 950 6772. E-mail: jhkwon@knu.ac.kr.

INTRODUCTION

Food irradiation has been proven as one of the alternative techniques to the existing methods for improving the hygienic quality, increasing the shelf-life, and enhancing the functional properties of different food items. Considering its various applications, especially those of a commercial nature, the technique can play a noteworthy role in food industry. Low doses of irradiation (<1 kGy) may be used to retard sprouting or rooting of bulbs, tubers and nuts during storage. High doses of irradiation (>10 kGy) have potential applications in developing sterilized food for special purposes. The irradiation alone, with a dose range of 1-10 kGy, or in combination with other effective technologies, can reduce the microbial load, kill the insect eggs and larva, affect the enzyme activity, and alter the sensory properties [Farkas and Mohancsi-Farkas, 2011].

The first commercial use of food irradiation in the world was carried out in Stuttgart in 1957 by Gewürzmüller and Co. [Maurer, 1958] to meet the consumer demand for spices with low bacterial count. However, the plant was shut down after a year in 1959 due to an amendment in the Food Law prohibiting food irradiation. Restriction was not due to any food safety issue, but due to a need for more research to check this new technology before commercial uses [Diehl, 1990]. In 1961, the major international organizations, such as FAO, IAEA and, WHO, recognized the usefulness of food irradiation and thus formed a coop-eration to discuss the necessary legal and safety aspects of irradiated foods. This had led to the establishment of a joint FAO/IAEA/WHO expert committee (JECFI: Joint Expert Committee on Food Irradiation), and during its first meeting in 1964, the importance of tests necessary to establish the wholesomeness of irradiated foods was determined [FAO, 1965]. In 1966, Professor J. F. Diehl was given the leadership of the Federal Research Centre for Nutrition, Nuclear Research Centre, Karlsruhe–Leopoldshafen, where work on the detection of food irradiation markers started. The institution had departments for food irradiation and food contamination, as well as one devoted in establishing the methods for radioactive tracers (indicators) in food materials [Ehlermann, 1999; IAEA, 1966]. Japan was the first country in Asia to start food irradiation in a commercial scale in 1973 with a potato irradiation plant in Shihoro Agricultural Co-operative, Hokkaido. This successful attempt got the attention of other countries [Loaharanu, 2003]. In 1980, JECFI had a meeting in Geneva and put a major landmark in their report in 1981 by confirming the safety of irradiated food. Consequently, the Codex Alimentarius Commission established the Codex General Standard for the Irradiated Food and the Recommended Code of Practice for the Operation of Radiation Facilities used for the Treatments of Foods. Due to the mandatory labeling requirements, the proper monitoring of irradiation process and traceability of irradiated product in the market the need for effective and reliable detection methods became more crucial. Thus serious efforts were made in this regard by the Research Co-ordination Program on Analytical Detection Methods for Irradiation Treatment of Foods (ADMIT) in late 1980s and early 1990s [Stewart, 2010].

The specificity and stability of radiation-induced changes are big constraints as the effects are usually similar to those produced naturally or through general food processing, such as heating and freezing [McMurray et al., 1996]. The European Union (EU) also played a leading role in the standardization of the available identification methods, conducting more than 30 interlaboratory blind trials to validate the proposed methods. The identification techniques can be categorized as physical, chemical, and biological methods depending on the basic principles of related techniques. The European Union has adopted framework Directive 1999/2/EC and implementing Directive 1999/3/EC for the uniform application of regulations regarding irradiated foods [Stewart, 2010].

Currently, there are 10 different standardized methods endorsed by the Codex Alimentarius (Table 1). The methods are classified as screening and confirmatory techniques depending on their specificity, reliability, reproducibility, and ease of use [Chauhan et al., 2009].

The general utility of food irradiation lacks consensus because of the negative perception of "nuclear technology". Some consumer groups also questioned the wholesomeness of irradiated foods on non-scientific grounds [Molins, 2001]. Different national regulations regarding irradiated foodstuffs are being implemented in various countries.

Table 1. European standards for the detection of irradiated foods which are also adopted by Codex

Standard No.	Method	Commodity	Validated products	Remark[1)]
EN1784	Gas chromatographic (GC) analysis of hydrocarbons	Food containing fat	chicken, pork, Beef, avocados, mangoes, papayas, camembert	Type II
EN1785	GC/MS analysis of 2-alkylcyclobutan-ones	Food containing fat	chicken, pork, liquid whole egg	Type III
EN1786	ESR spectroscopy of bones	Food containing bone	chicken, fish, frog legs	Type II
EN1787	ESR spectroscopy of cellulose	Food containing cellulose	paprika powder, pistachio nut shells, strawberries	Type II
EN1788	Thermoluminescence of silicate minerals	Food containing silicate minerals	herbs and spices, shrimps	Type II
EN13708	ESR spectroscopy of crystalline sugars	Food containing crystalline sugar	dried papayas, dried mangoes, dried figs, raisins	Type II
EN13751	Photo-stimulated luminescence	Food containing silicate minerals	herbs and spices, shellfish	Type III
EN13783	Microbial screening using direct epifluorescent filter technique/aerobic plate count (DEFT/APC)	Herbs, spices and raw minced meat	herb and spices	Type III
EN13784	DNA comet assay screening	Food containing DNA	chicken, pork, plant cells, e.g. seeds	Type III
EN14569	Microbiological screening for using LAL/GNB procedure		chicken, pork, beef	

[1)] Type II: reference methods; Type III: alternate approved methods.

Nevertheless, the technical effectiveness of food irradiation for community health and safety has been extensively investigated and well-endorsed by the international health authorities. Currently, more than 55 countries have regulated the use of irradiation for different food items in various categories, where more than 30 countries have irradiated food products in their local markets [Farkas and Mohancsi-Farkas, 2011]. To safeguard the consumer and create awareness of this science-based technology, proper labeling of irradiated food for distribution is mandatory. Regardless of the amount of the irradiated ingredient, each food must be labeled "irradiated" or "treated with ionizing radiation" with the optional use of "Radura" mark (Figure 1). This applies to both pre–packed and loose (in bulk) foodstuffs [CODEX STAN 1-1985, Amend. 7-2010]. There is a great potential for the export of these irradiated products to the countries where irradiation is not allowed or which have different well-defined strict regulations related to the trade of irradiated food products. One obstacle hindering the acceptance of food irradiation is the difficulty of determining whether or not the food has been irradiated. In this regard, proficient detection methods to classify marketed foods as irradiated or non-irradiated are needed [Arvanitoyannis, 2010]. The validated and effective detection techniques are also required for the enforcement of related regulations and fulfillment of quarantine requirements in order to facilitate trade between different countries [Chauhan et al., 2009].

Although European nations were the pioneer for the commercialization of food irradiation, a decreasing trend in the use of irradiated foods has been observed in Europe due to the strict regulations [Kume et al., 2009]. Elahi et al. [2008] reported the first court judgment in Europe against a spice supplier on the basis of laboratory reports confirming positive results for the irradiation status of samples. There is no general method valid for all types of food, nor is one expected in the near future, because of the dynamic nature of food items. The effects of long-term storage and adverse processing conditions on radiation-induced markers are also important issues regarding the identification of irradiated food [Ahn et al., 2012a; 2012b].

Detection of irradiated ingredients added in low proportion in a complex formula food may also be a big challenge for a food analyst [Ahn et al., 2012c]. McMurray et al. [1996] provided technical and practical criteria for an ideal identification method. Generally, irradiation produces changes that are identical to those from other food processing techniques and the quantity of radiation-induced markers is too low to detect [Chauhan et al., 2009].

Figure 1. The mark named as "Radura" to demonstrate the irradiated food.

With the compliance of related regulations, particularly the labeling requirement, identification methods are also needed to confirm the good irradiation practices and establish traceability and documentation system. Generally, a specific irradiation dose is needed to achieve the desired technological benefits where an over-dose may result in detrimental quality changes (e.g. off-odor, softening in physical texture, color change) [Akram et al., 2012c]. In this case, dose-estimation could be very helpful for the precise application of the technology [D'Oca et al., 2009].

Different advanced countries importing agricultural commodities require food irradiation as a quarantine treatment, where identification methods could facilitate the process of clearance in accordance to the applied regulations [Arvanitoyannis, 2010; Marchioni, 2006]. The potential identification methods are described in the following sections with respect to their basic principle, application range, specificity, and reliability.

PHOTOSTIMULATED LUMINESCENCE (PSL) ANALYSIS

Mineral dusts, especially silicate materials, present on the food can store energy upon irradiation in charge carriers trapped at structural or interstitial sites. The external light stimulation of these minerals could release the stored energy, giving a measurable photon count [EN 13751, 2009]. The phenomenon of luminescence is illustrated in Figure 2.

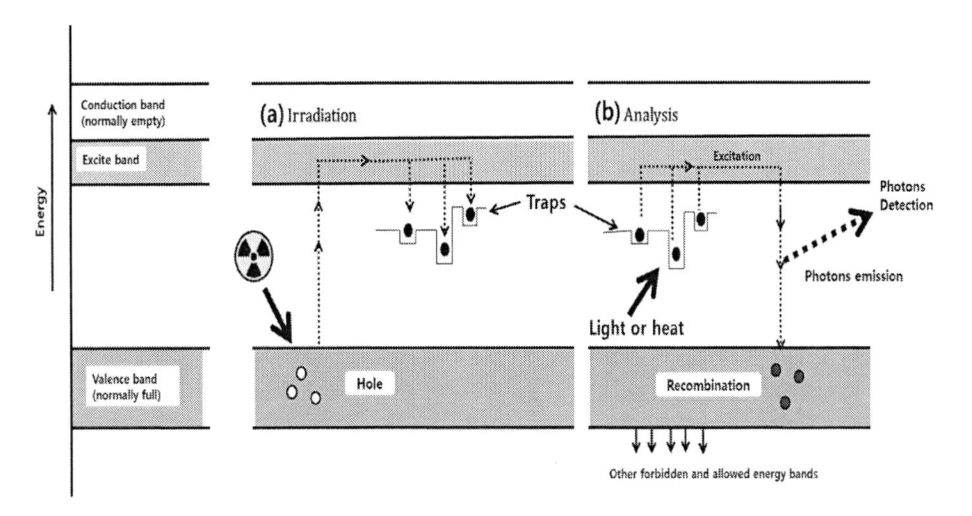

Figure 2. Principle of photo- and thermo-luminescence of silicate minerals separated from irradiated food materials [Tsoulfanidis, 1995].

The successful application of pulsed-photostimulated luminescence to screen irradiated spices was initially reported by Sanderson and others [1993], demonstrating the effectiveness and simplicity of PSL analysis. This technique does not require any laborious separation and/or preparation steps, providing a very simple and rapid analysis procedure [EN 13751, 2009]. The whole sample is uniformly spread in a small petri dish (50 mm, n=2) and may be cut to fit in the dish. The samples are loaded in sample chamber and the PSL signals are recorded for 60 sec. The results are given in the form of accumulated PSL count per 60 sec.

The lower (700) and upper threshold (5000) limits, which are primarily validated using herb and spices (different for shellfish, 1000-4000 count per 60 sec), are used to interpret the PSL results (Figure 3).

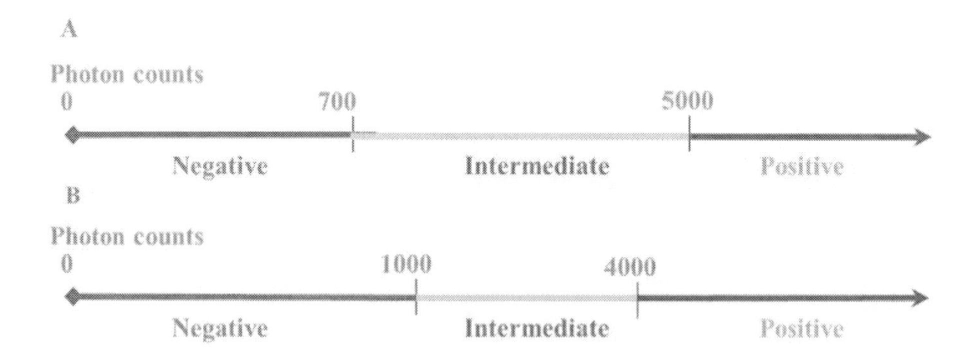

Figure 3. Threshold value of PSL photon counts. (A, spices and herbs; B, shellfish).

The samples with PSL count less than the lower threshold limit are classified as negative, while those with PSL count higher than the upper threshold limit are classified as positive. The PSL results between these two limits are regarded as intermediate that need further investigation to confirm their status. The PSL method is acceptable as a screening approach only as it may lead to wrong classifications of the samples because of the signal bleaching (false-negative) and cross-contamination (false-positive) [Bortolin et al., 2007; Ahn et al. 2012b] . Recently, we reported that long-term storage under different light conditions resulted in a time and light condition-dependent bleaching of PSL signals [Ahn et al, 2012b]. Larger mineral content or high salt concentration could also produce false (positive) results [Fuochi et al., 2008; EN 13751:2009]. The protocol EN 13751:2009 recommends the calibration measurement after screening analysis, in which the same sample is irradiated again using a defined irradiation dose and the PSL count is re-measured. Irradiated samples exhibit small increase, while non-irradiated ones show substantial elevation, in the PSL count. This step also confirms the sensitivity and availability of mineral debris to store energy upon irradiation.

Thermoluminescence (TL) Analysis

TL analysis is considered as the most promising method for the identification of irradiated food materials from which silicate minerals (contaminating dusts) could be separated [Chauhan et al., 2009]. Along with TL, chemiluminescence (CL) was also studied as a possible detection technique for irradiated foods, where alkaline luminolhaemin solution is added in the sample to produce luminescence that is determined using a light detector. However, due to some inherent limitations of CL, it is now considered as an inappropriate identification method for irradiated food [Molins, 2001]. Upon irradiation, the mineral fraction of foodstuff stores energy through a charge trapping process. This energy could be released through a controlled heating (Figure 2), resulting in a measurable luminescence that produces a characteristic TL glow curve. The EN 1788 European standard is applicable to all food materials having enough (about 0.2 mg) silicate minerals. The TL properties principally depend on the quality and quantity of the separated silicate minerals regardless of the type of food [Beneitez et al., 1994; Calderon et al., 1995; Gastelum et al., 2002].

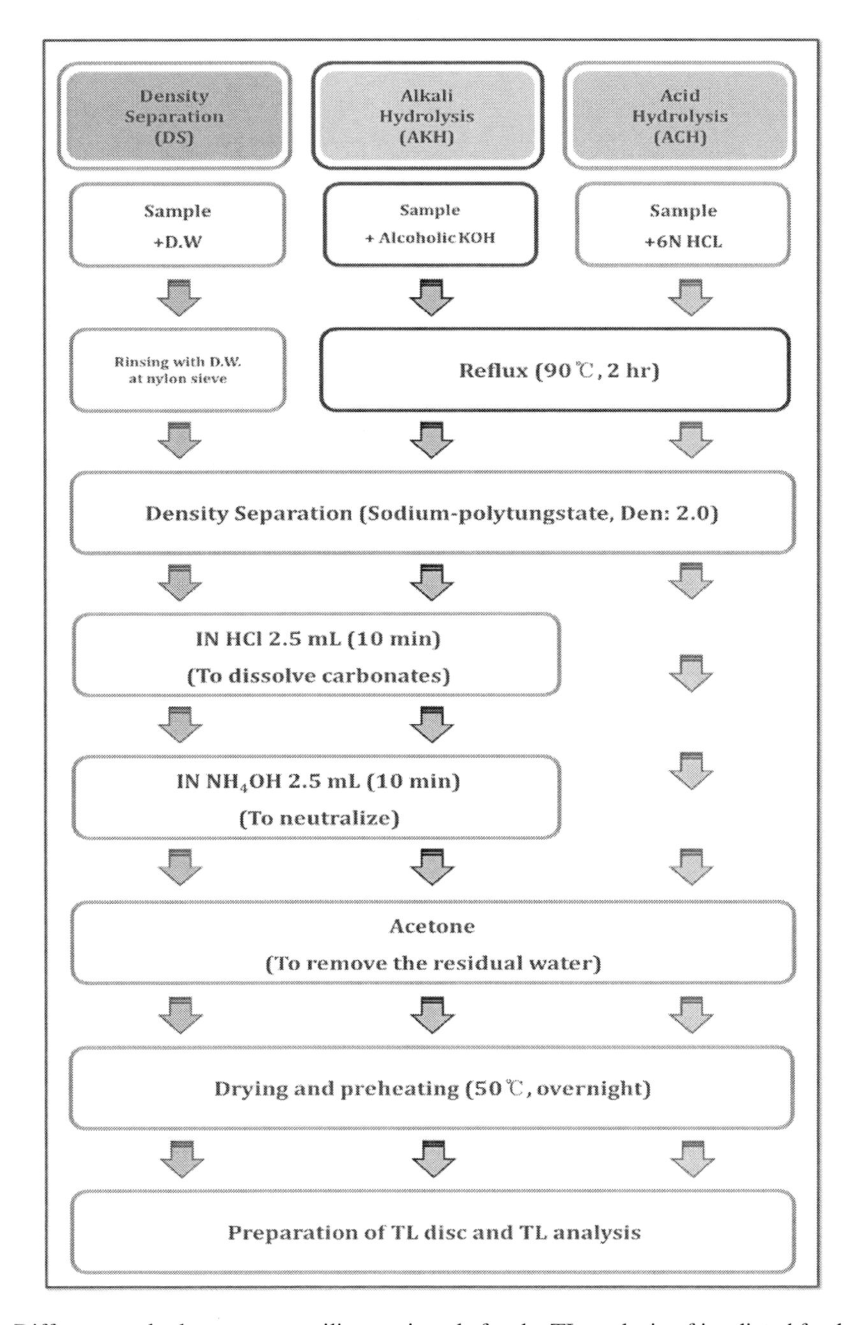

Figure 4. Different methods to separate silicate minerals for the TL analysis of irradiated foods.

The silicate minerals usually exhibit high sensitivity to applied radiations providing high TL intensity with good stability of the TL signals during storage of the material [McKeever, 1985].

While different methods may be adopted to separate the mineral fractions (Figure 4), the typical method usually involves separation of inorganic minerals (density over 2.5 g/cc) through washing of the food sample with distilled water and purification from organic matter (density less than 1.1 g/cc) through a density separation approach using heavy liquids, such as sodium polytungstate or carbon tetrachloride.

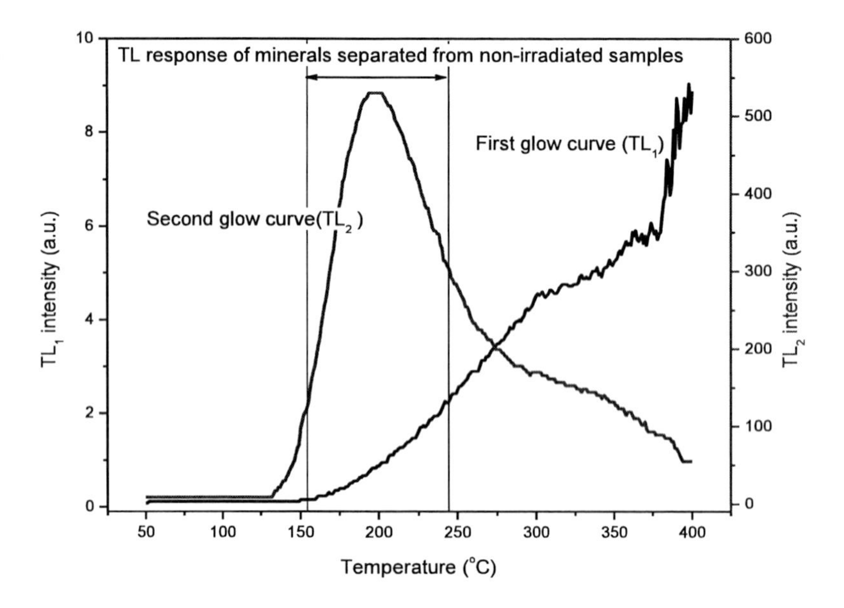

Figure 5. TL glow curve of silicate minerals separated from non-irradiated samples.

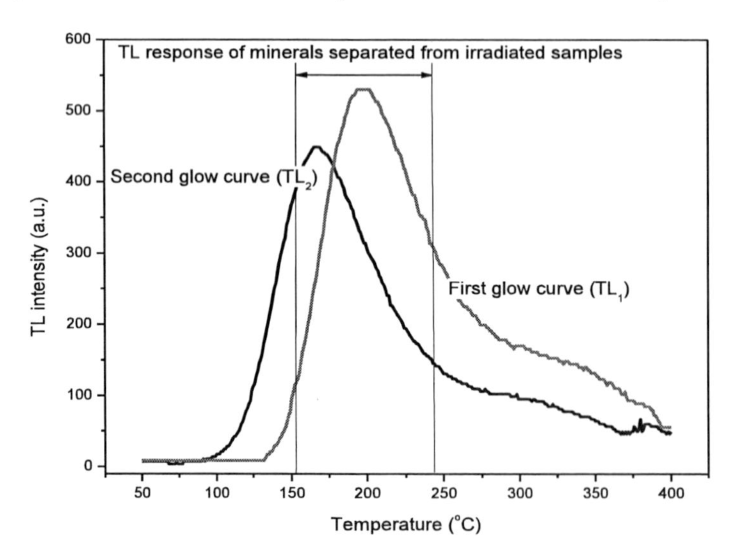

Figure 6. TL glow curve of silicate minerals separated from irradiated samples.

The separated minerals are measured using a TLD reader (also used for age-dating studies), applying a controlled heating (0-400°C at the rate of 6°C/Sec) under a nitrogen atmosphere. During analysis, the TL signals are recorded as a function of temperature and the resulting curve is called a TL glow curve (Figures 5 and 6). The glow curve intensity and shape are the typical characteristics used to interpret the TL results, confirming the irradiated status of samples. Although the TL analysis provides an excellent qualitative discrimination of irradiated and non-irradiated samples, accurate quantitative results are not easy to obtain due to variable mineralogical compositions and their related TL response. Long-term storage and processing can also change TL behaviors making it difficult to estimate the applied dose [D'Oca et al., 2009].

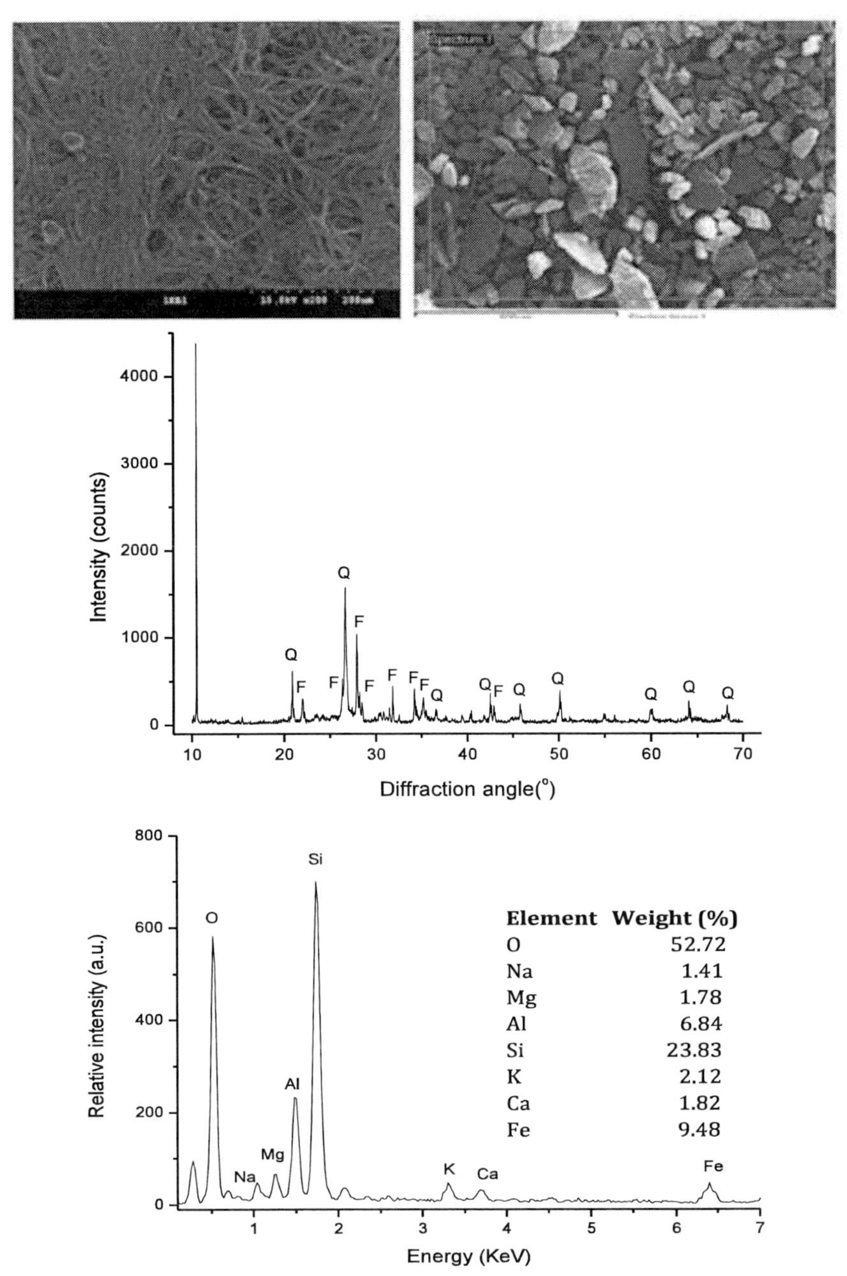

Figure 7. The inorganic minerals present upon samples (above) were separated and characterized using XRD (middle; mineral composition: Q, Quartz & F, Feldspar) and EDX (below; elemental composition) spectroscopy [Akram et al. 2012].

The detectionability and stability of the analysis will depend on both the quality and quantity of the separated minerals and the TL glow temperature range selected for the interpretation of results. In this regard, the detailed composition of the separated minerals (usually analyzed through X-ray diffraction or energy-dispersive X-ray spectroscopy) could be helpful to understand the TL response of the separated minerals (Figure 7). However, variable results may be found from similar mineralogical composition depending on the mineral origins (pegmatite, volcanic or hydrothermal) and their sensitivities to the absorbed dose [McKeever, 1985]. Different natural, processing, and environmental conditions can also change the mineral sensitivity and TL response [Garcia-Guinea et al., 2007]. Feldspar and quartz are the main silicate minerals important for TL analysis to confirm the irradiation status of food materials. Soika and Delincée [2000] reported the outstanding sensitivity of feldspar minerals compared with the others, which define the overall TL glow curve characteristics even if present in relatively low concentration in the separated mineral fraction.

Considering the variable TL responses of separated minerals, the analytical method involves a normalization step. The minerals on the TL discs, already measured for the first TL glow (TL_1), are again irradiated (usually 1 kGy as recommended by EN 1788, 2001) and measured for the second TL glow curve (TL_2). The TL glow ratio (TL_1/TL_2) is used to confirm the TL results (Figures 5 and 6) since the irradiated samples exhibit higher TL ratios than the non-irradiated ones. The TL ratios also depend on the integrated TL glow temperature range selected for the analysis. Generally, a temperature range of 150-250°C is used for the calculation of TL ratio. However, the protocol EN 1788 [2001] recommends the calibration of the TL temperature scale using well-characterized standard materials like LiF TLD-100® to get a more precise result (Figure 8).

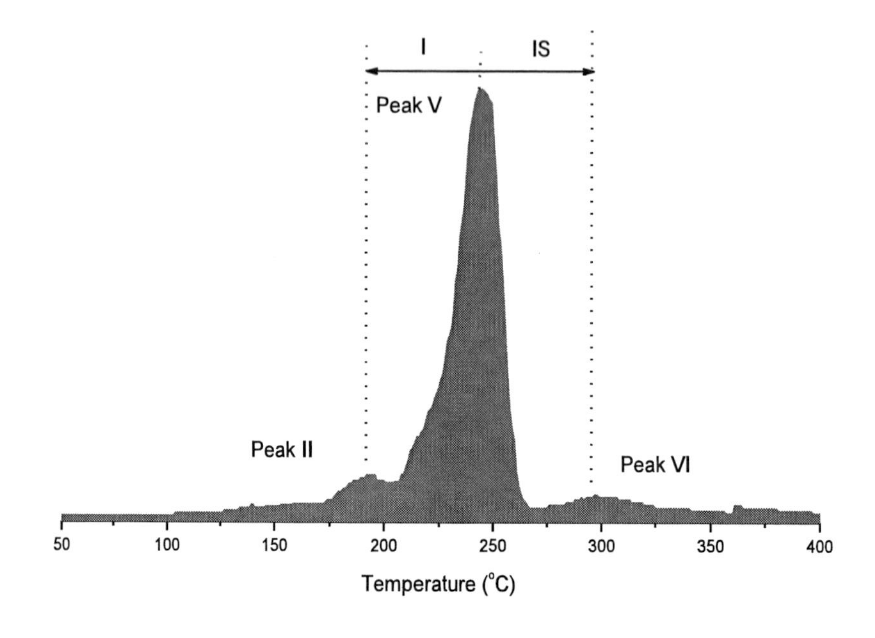

Figure 8. TL glow curve of 0.5 Gy-irradiated LiF with marked temperature intervals in accordance to EN 1788:2001 Annex B.

The separated minerals from non-irradiated samples show a residual geologically-derived TL signals with maximum peak intensity at/after 300°C and TL ratio of less than 0.1. Irradiation induces typical TL glow characteristics, particularly the TL intensity and TL glow curve shape. Silicate minerals from irradiated samples usually exhibit high intensity TL glow curve with maximum peak at a temperature range of 150-250°C. The TL ratio should also be greater than 0.5. In the case of a TL ratio between 0.1 and 0.5, the first TL glow curve provides the conclusive information [EN 1788, 2001].

For complex food products containing irradiated ingredients [Ahn et al., 2012c] or heat-processed irradiated food materials [Lee et al., 2008a; Kim et al., 2012; Ahn et al., 2012d], the TL ratio could not be used as a confirmatory approach, but the careful interpretation of the TL glow curve could provide the required information for making a decision about the irradiation status of the sample. When interpreting the TL results, a scientist should consider anomalous fading that could be mainly observed in feldspars [Li and Li, 2008], dose-saturation depending on the inherit characteristics of minerals [Duller et al., 2003], samples containing natural radionuclides (^{40}K, ^{238}U, ^{232}Th, etc.) resulting in self-irradiation [Vandenberghe et al., 2008], and history of minerals (heat-processed) leading to changes in TL characteristics. In a practical processing environment, food irradiation is usually combined with other techniques that can affect the TL results [Ahn et al., 2012d]. The type and rate of irradiation can also affect the key characteristics of the TL glow curve [Soika and Delincée 2000]. In addition, storage conditions (temperature, presence or absence of daylight, etc.) and time can affect the TL glow curve by shifting the maximum peak of the TL glow curve towards the higher temperature. We have shown the significant effect of two-year storage under different light conditions on the TL identification properties, suggesting that identify-cation was still possible [Ahn et al., 2012b].

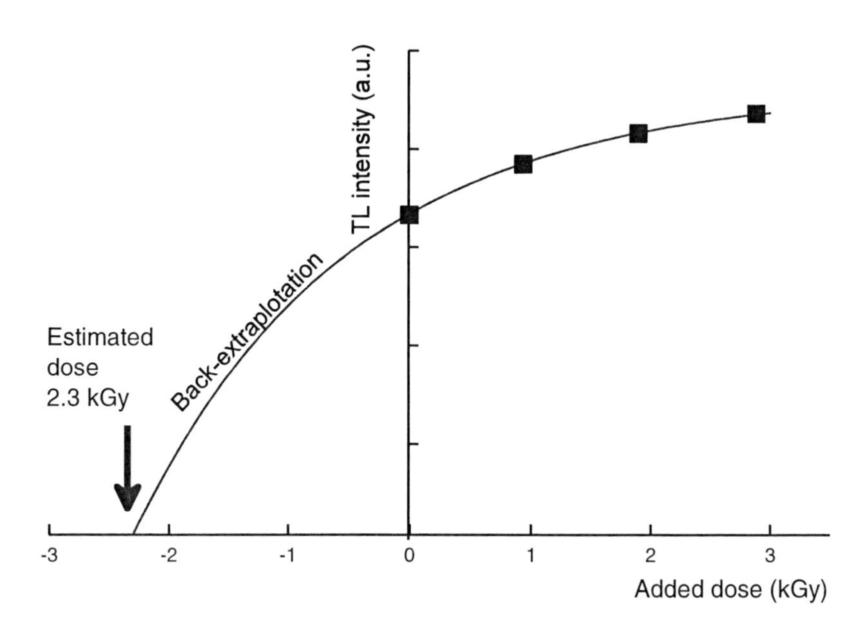

Figure 9. The TL-based methodology of dose estimation through additive-dose and back-extrapolation procedure [D'Oca et al., 2009].

Variable quantitative response and mineral composition are the main constraints for the development of TL dose-estimation approach. D'Oca et al. [2007; 2009; 2010] has tried the TL approach for the estimation of applied dose through additive-dose procedure. After measuring the original TL glow curve (TL_1), the samples were re-irradiated several times and analyzed for TL_2, TL_3, and TL_4 intensities. Since the TL analysis removed the TL signals, the results from the re-irradiation steps were corrected by adding the TL_1 intensity. The TL_1, TL_2, and TL_3, and TL_4 intensities were used to estimate the absorbed doses using linear and polynomial expressions and back-extrapolation method (Figure 9) [Stewart and Stevenson, 1997; D'Oca et al., 2009]. The results could be used to develop an in-house approach to confirm the applied dose using an irradiated product. However, a more detailed investigation is needed involving the application of this approach in different irradiation detection laboratories for a quantitative TL analysis of irradiated food.

Electron Spin Resonance (ESR) Spectroscopy

Electron spin resonance/electron paramagnetic resonance spectroscopy detects unpaired electrons present in the samples due to the defects in semiconductors, paramagnetic ions derived from transition or main group elements, and free radicals (may or may not be radiation-induced). External magnetic field is applied to produce an energy level difference of the electron spins $m_s = +½$ and $m_s = -½$. Microwave beam is then applied and the resonance absorption is measured. Conventionally, ESR spectrum illustrates the first derivative of the absorption against the applied magnetic field [EN 1787, 2000]. Ionizing radiations produce free radicals that are generally very short-lived depending on the sample matrix. The ESR analysis could be affective if the generated radicals have enough life that is comparable to the shelf-life of the food product. Generally, the dried components with radicals that have limited activity due to the rigid structure of the matrix could be used. The specificity of the signals is also important as radicals may also be induced through various processes commonly applied in food industry [Desrosiers, 1996; EN 1787, 2000]. The European Union has standardized three ESR methods mainly targeting food containing bone [EN 1786, 1997], cellulose [EN 1787, 2000] and crystalline sugar [EN 13708, 2002].

Radiation-Induced ESR Signals in Food Materials of Plant Origin

In plant foods, irradiation can induce free radicals in cellulose and crystalline sugars and the unpaired electrons in the radicals could be used to determine the irradiation history of the samples [Kwon et al., 2000]. The European Union has standardized two ESR methods for the identification of irradiated foods, such as dried spices and dried fruits, targeting the radiation-induced cellulose [EN 1787, 2000] and crystalline sugar radicals [EN 13708, 2002], respectively. In fresh fruits and vegetables with higher moisture contents, the radicals produced upon irradiation are not as stable as those from dried food materials. Moreover, sugar does not exist in crystalline form, so formation of crystalline sugar radicals is not possible. In general, the hard parts (seeds, shell, skin etc.) of fresh food materials are used for ESR-based detection of cellulose radicals where different drying techniques are also used to reduce the moisture content, thus resulting in a clear ESR signal [de Jesus et al., 2000; Jo and Kwon, 2006]. For identification purposes, the first ESR study of strawberry seeds was reported by Raffi and others [1988], demonstrating the dose-dependent effect of irradiation on the ESR signal, which was also sensitive to the water contents of the samples. Various scientists reported promising results on the ESR-based identification of irradiated herbs, nuts, spices, fruits, and vegetables [de Jesus et al., 2000; Kwon et al., 2000; Jo and Kwon, 2006; Chauhan et al., 2009]. The non-irradiated samples may exhibit silent ESR signals (Figure 10A), but a central ESR signal (due to natural organic radicals) is usually observed in food materials of plant origin, regardless of the irradiation history (Figure 10B). The naturally-induced semiquinone radicals, produced by the oxidation of polyphenolic compounds in the plant matrix, are considered mainly responsible for this central ESR signal [Calucci et al., 2003].

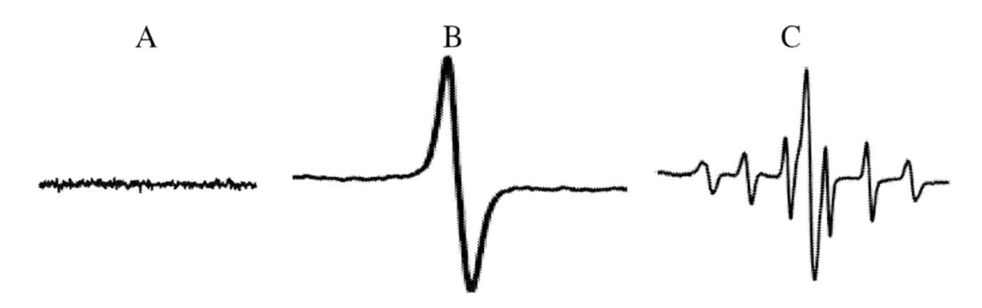

Figure 10. Typical ESR spectra of non-irradiated foods (A, ESR silent (no signal); B, natural signal usually in food of plant origin; C, the effect of Mn^{2+}).

In addition, a sextet signal attributed to Mn^{2+} could also be observed (Figure 10C), making the detection difficult due to it's overlapping with the radiation-induced signals [Akram et al., 2012b]. Upon irradiation, the intensity of the central signal increases with the emergence of two side peaks that are usually attributed to radiation-induced cellulose radicals (Figure 11). The 6 mT mutual spacing of these two side-peaks is radiation-specific [EN 1787, 2000], however, the absence of this signal does not confirm the absence of irradiation treatment since the ESR results depend on a number of factors, such as irradiation dose, sample composition, conditions (especially temperature) during irradiation, and storage conditions [Stevenson and Gray, 1995; Lee et al., 2002]. Deighton et al. [1993] accounted that only the left-side peak is due to the radiation-induced cellulose radicals, whereas the right -peak signal, which is also sensitive to heat processing [de Jesus et al., 1996], is due to the lignin radicals. Ukai [2004] reported that the ESR signal in black pepper may be affected by the presence of transition metal ions, such as Fe^{3+} and Mn^{2+}. However, due to the different positions of signals from these paramagnetic ions and spacing between two manganese peaks (about 8-9 mT), the identification of radiation-specific signals is possible.

De Jesus et al. [1999] reported the successful application of an alcoholic-extraction technique to reduce the moisture content in fresh kiwi, papaya, and tomato. A similar technique was also used by Delincée and Soika [2002], showing improved ESR signal from cellulose radicals in strawberry, papaya, and some spices. Jo and Kwon [2006] used oven drying at 50°C for the determination of cellulose radicals in irradiated kiwi fruits. For irradiated fresh mangoes, Kikuchi et al. [2010] have reported the detectionability of radiation-specific ESR signal after one week of post-irradiation storage.

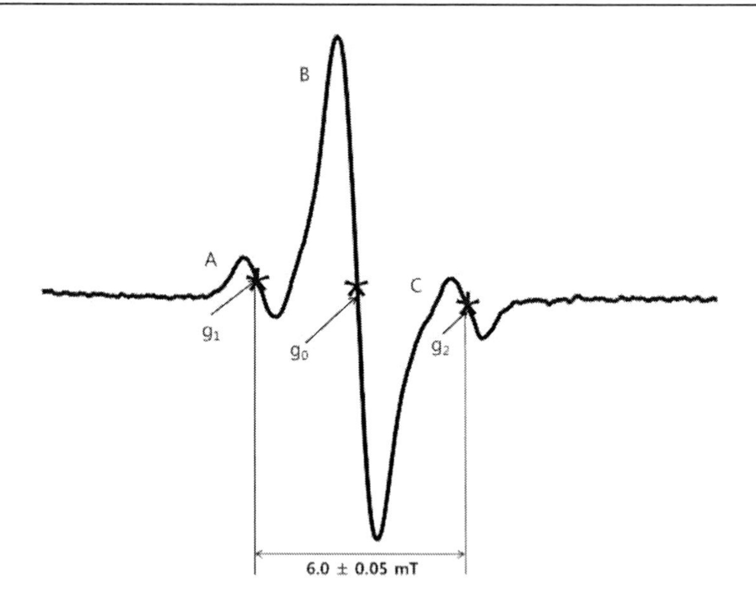

Figure 11. Typical ESR spectrum of irradiated cellulose-containing foods.

Different drying techniques, such as alcoholic-extraction, freeze-drying, oven-drying, and air-drying were conducted to compare ESR-based identification results in irradiated fresh fruits, demonstrating the effectiveness and convenience (short time and cheap) of the alcoholic-extraction technique [Yordanov et al., 2006; Aleksieva et al., 2009; Yordanov and Aleksieva, 2009]. An alternative technique involving the measurement of ESR signal in irradiated fresh papaya under liquid nitrogen environment (77 K) was also proposed by Kikuchi et al. [2011]. Dried fruits and vegetables usually contain crystalline sugars which could provide multicomponent ESR signals following irradiation (Figure 12). Various ESR signals are possible due to the difference in the composition and crystallinity of the mono- and disaccharides in the sample [EN 13708, 2002].

Radiation-specific crystalline sugar radicals were found stable over several months of storage providing enough information to characterize the irradiated samples [Malec-Czechowska et al. 2003]. The sensitivity of samples to applied-radiation doses could be confirmed by irradiating a small part of the sample and re-analyzing for ESR signals. ESR signals due to crystalline-sugar radicals induced by irradiation were observed in irradiated dried fruits and vegetables [Desrosier and Mclaughlin 1989; Kwon et al. 2000] and complex-formula foods containing sugar [Ahn et al. 2009 a,b]. Malec-Czechowska et al. [2003] successfully characterized different mushroom varieties using ESR spectroscopy with the exception of irradiated dried *Lentinus edodes*.

Recently, our results showed that ESR-based identification was possible even at a low dose of 2 kGy when the cap skin or core of irradiated dried *Lentinus edodes* was used for the ESR analysis [Akram et al., 2012a]. Noteworthy results were also reported by Karakirova et al. [2010] on the dosimetric properties of different gamma-irradiated sugar samples.

Qualitative results are usually reported for the discrimination of non-irradiated and irradiated samples. However, some scientists have tried to develop a dose-estimation approach using re-irradiation technique and the results could be useful for an in-house monitoring of samples with least variations in the sample composition [Kwon et al., 2000].

Recently, scientists also reported the similarity of grinding-induced ESR signals to those of low-dose (10-50 Gy) irradiated pure lactose [Lee et al., 2010] and sucrose samples [Ahn et al., 2012e]. Although the specificity of the radiation-induced sugar signals was challenged, the intensity of the grinding-induced ESR signals was too low to cause any considerable problem in the identification of irradiated foods [Ahn et al., 2012e].

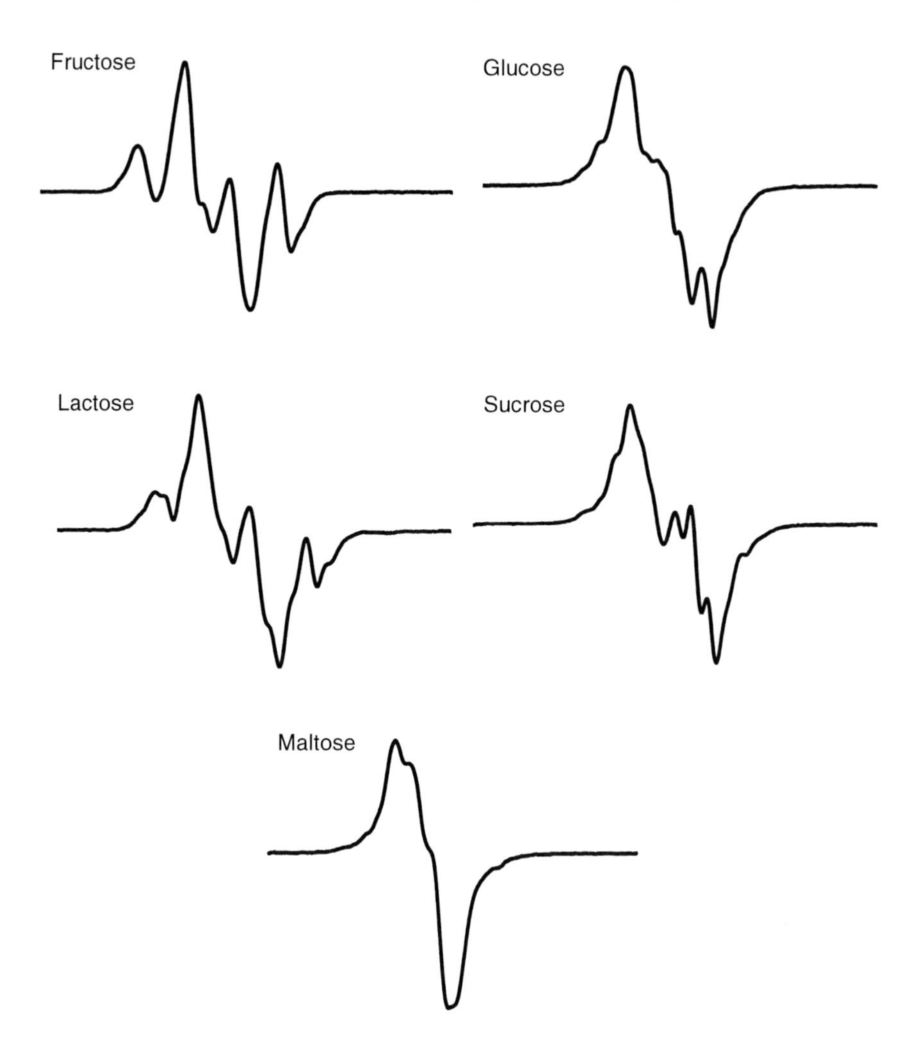

Figure 12. Typical ESR spectra of irradiated crystalline sugars.

Radiation-Induced ESR Signals in Food Materials of Animal Origin

The ESR signals in irradiated bone samples were first reported by Gordy et al. [1955] and were suggested as potential irradiation makers by Onderdelinden and Strackee [1974]. Reported scientific data have shown the possibility of identification of almost all kinds of irradiated meat containing bones, including chicken, frog, lamb, turkey, pork, beef, goose, duck, and rabbit [EN1786, 1997; Chauhan et al., 2009]. Non-irradiated bone samples usually exhibit silent (straight line) for ESR signals, however, a single peak (g=2.0044) attributed to the bone-marrow of samples could be found. Upon irradiation, two types of signals are usually observed: i) from bone collagen and ii) due to structural defects in bone hydroxyapatite ($Ca_{10}(PO_4)_6(OH)_2$) [Ostrowski et al., 1974; Sin et al., 2005]. The radicals induced in the hydroxyapatite following irradiation generate an asymmetric absorption at g~2.0030–2.0033 and g~1.9969–1.9975 (Figure 13). Qualitative characteristics of ESR signals produced from all irradiated bones were similar and mainly due to CO_3^{1-}, CO_3^{3-}, and CO_2^{1-} ions produced upon irradiation and trapped in complex bone matrix [Serway and Marshall, 1967; Cevec et al., 1972; Gray and Stevenson, 1989].

The radiation response of different bone samples against applied doses was found linear showing a good increase in ESR intensity with the increase in applied-radiation dose [Chawla and Thomas, 2004]. Variation in ESR signal intensity was reported to be dependent on the age of animal and type of bone due to the differences in the crystallinity of the bone materials [Ostrowski et al., 1980; Glimcher, 1984; Gray et al., 1990]. These radiation-induced signals are extremely stable at room temperature even after years of storage and against different heat-processing conditions [Bhatti et al., 2012]. Marchioni et al. [2002] also reported the good ESR results after extraction (enzymatic hydrolysis) of bone fragments from mechanically recovered meat.

Comparable results were found in fish bone samples. The ESR technique has been successfully applied to trout, sardine, salmon, whiting, halibut, cod and mackerel fish bone samples and the results were reproducible during storage. Better results were reported when fish teeth were used as a sample [Raffi and stocker, 1996; Empis et al., 1995; Chiaravalle et al., 2010]. In a comparative study, pork bone showed more promising results than salmon bone samples mainly due to the higher crystalline structure of hydroxyapatite in pork bones [Goodman et al., 1989]. The exoskeleton or shell of irradiated crustaceans could provide the evidence of irradiation history through ESR analysis [Dodd et al., 1985; Goodman et al., 1989], where the aragonite or calcite minerals are responsible for the radiation-specific ESR signals [Desrosiers, 1989; Stewart et al., 1994; Bhatti et al., 2012]. The species and geographical origins of prawn and shrimp were also found to be important factors affecting the radiation-induced ESR signals [Stewart and Kilpatrick, 1997]. Hence, the chemical composition (mainly aragonite and calcite minerals) of the shell is considered important for radiation-specific ESR signals.

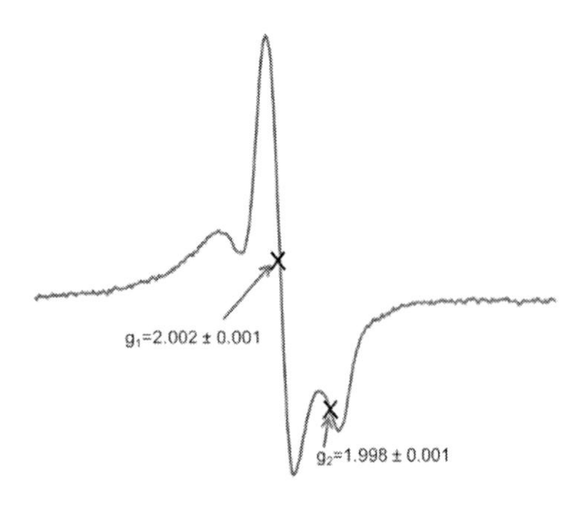

Figure 13. Typical ESR spectrum of irradiated bone.

RADIATION-INDUCED CHEMICAL MARKERS

Food irradiation improves the shelf-life and microbiological quality of different food materials without compromising their nutritional safety [WHO, 1999]. Radiation-induced changes are mostly similar to those of the other routine preservation techniques [Diehl, 1995]. Ionizing radiations could have a direct or indirect action on the food components through reactive intermediates (free radicals) that mainly produced by the radiolysis of water molecules [Diehl, 1995]. The electron-deficient carbon-carbon double bonds of unsaturated fatty acids and carbonyl groups (fatty acids and amino acids) are particularly susceptible to the radiation-induced radical species [Thakur and Singh, 1994]. The radiation-induced hydroxyl radicals in aqueous [Thakur and Singh, 1994] or in oil emulsion food systems **[O'Connell and Garner 1983]** could target the acyl-oxygen bond-producing aldehydes, C_{n-1} alkanes, short-chain hydrocarbons, CO, free fatty acids, and alcohols [Josephson and Peterson, 2000].

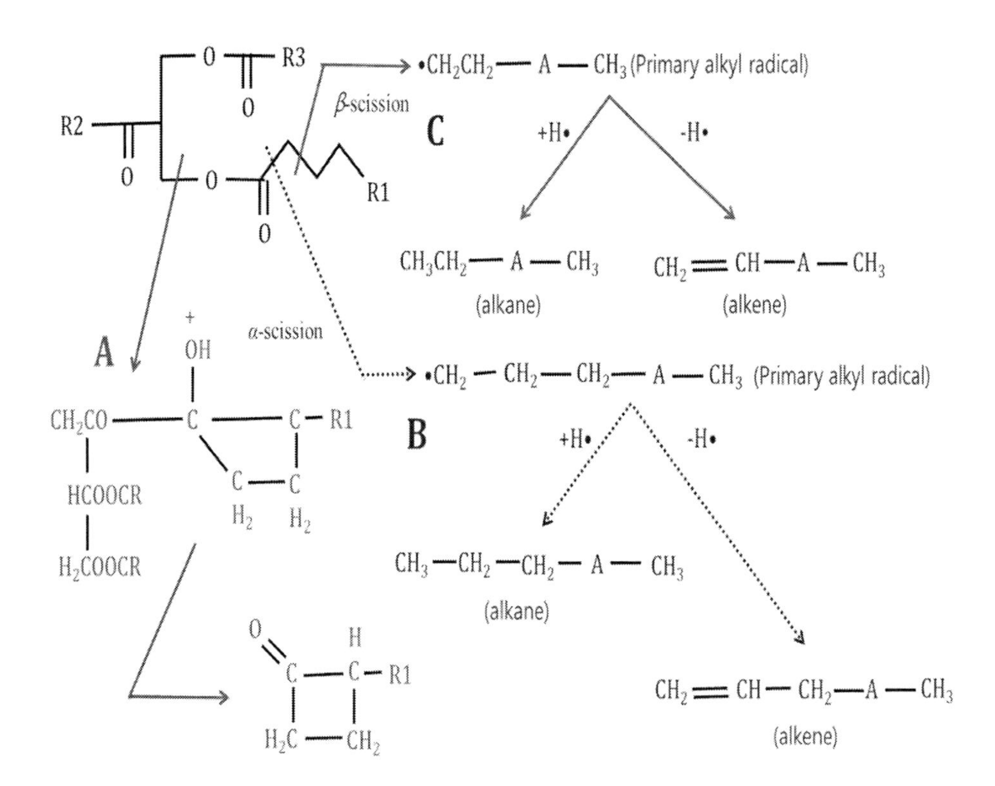

Figure 14. Radiation-induced cleavage of triglyceride molecules resulting in 2-alkylcyclobutanone (A) and hydrocarbons (B and C) [Marchioni, 2006].

The radiolytic products of lipids and lipid-containing foods mainly depend upon the fatty acid composition of foods [LeTellier and Nawar 1972]. A number of scientists [Schreiber et al., 1994; Delincée, 2002; EN 1784, 2003] reported the radiation-induced hydrocarbons and 2-alkylcyclobutanones (Figure 14) as detection markers in characterizing the irradiation history of fat-containing foods, where the concentrations of these major radiolytic compounds show a linear relationship with the irradiation dose and temperature during the processing.

The stability of these radiation-induced chemical changes is usually time-dependent and significant affected by post-irradiation storage and processing conditions [Kim et al., 2009; Kwon et al., 2012].

Radiation-Induced Hydrocarbons

The radiation-induced hydrocarbons among the volatile compounds in irradiated food products were first reported in 1959 by Merritt and co-workers, in which they suggested the specificity of these hydrocarbons to radiation treatment only [Merritt et al., 1965]. Champaign and Nawar [1969] also reported the radiation-induced hydrocarbons as major radiolytic products in fat systems, where their patterns depended on the fatty acid composition of the food material. The chemical C–C bonds in the triacylglycerol molecules generally split up at the α- and β-positions (regarding the carbonyl group) upon irradiation (Figure 13). Two types of hydrocarbons are mainly produced from fatty acids: (i) hydrocarbons lacking one carbon atom than the parent fatty acids (C_{n-1}) and (ii) hydrocarbons lacking two carbon atoms with an additional double bond at position 1 ($C_{n-2,\ 1-ene}$) [Spiegelberg et al., 1994]. Consequently, the 8-heptadecene ($C_{17:1}$) and 1,7-hexadecadiene ($C_{16:2}$) from oleic acid, n-pentadecane ($C_{15:0}$) and 1-tetradecene ($C_{14:1}$) from palmitic acid, and n-heptadecane ($C_{17:0}$) and 1-hexadecene ($C_{16:1}$) from stearic acid are usually predictable in food materials containing fat. The occurrence of radiation-induced saturated hydrocarbons with odd number of carbon atoms (C_{n-1}) and unsaturated hydrocarbons ($C_{n-2,\ 1-ene}$) could be used as irradiation detection-markers in fat-containing foods [Kwon et al., 2012]. The pattern of these radiation-induced hydrocarbons could also be important for the correct identification since the hydrocarbons $C_{17:1}$ and $C_{16:2}$ generated from oleic acid (most abundant fatty acid in food materials) are usually found in higher concentrations than any other pairs of hydrocarbons from a fatty acid [Hwang, 1999].

The European standard EN 1784 [2003] is based on the radiation-induced hydrocarbons, which are examined by gas chromatography. The analysis of radiolytic hydrocarbons involves extraction of fat, isolation of the hydrocarbon fraction by adsorption chromatography (Florisil), and determination of hydrocarbons using gas chromatography (coupled with a flame ionization detector or a mass spectrometer). The interlaboratory trials were also conducted, resulting in a successful validation of this method for the identification of irradiated raw chicken, pork, beef, Camembert cheese, avocado, papaya, and mango samples. In order to minimize the long-extraction time, alternative/modified analytical techniques, such as solid-phase microextraction (SPME), were used to identify the radiolytic markers in meat [Fan and Sommers, 2006] and meat products [Kim et al., 2005]. The microwave-assisted extraction of fat for hydrocarbon analysis could also enhance the efficiency of this detection approach [Kwon et al., 2007].

Recently, the successful identification of radiation-induced hydrocarbons in irradiated pork, beef, and chicken before and after cooking was reported by Kwon et al. [2012; 2011a,b]. Similar results were also observed by Lee et al. [2008b] in gamma-irradiated sesame seeds after steaming, roasting, and oil extraction. Nam et al. [2011] presented valuable findings regarding the dose-dependency of radiation-induced hydrocarbons in e-beam irradiated vacuum-packaged sausages.

Radiation-Induced 2-Alkylcyclobutanones (2-ACBs)

The formation of 2-ACBs in irradiated lipid samples was first reported in 1972 [Le Tellier and Nawar,1972]. However, it was only in the 1990s that the use of radiation-induced 2-ACBs as a detection-marker was proposed [Raffi et al., 1993]. Consequently, advanced analytical techniques and new methods for the isolation and detection of 2-ACBs were developed through time [Crews et al., 2012]. The 2-ACBs consist of four-membered ring with a keto group at position 1 and a side chain at position 2 (Figure 13). As mentioned earlier, in the case of radiation-induced hydrocarbons, the acyl-oxygen bond in acylglycerols is sliced upon irradiation, resulting in the formation of alkanes, alkenes, lactones, ketones, esters, aldehydes, and 2-ACBs [Stewart, 2001]. The 2-ACBs have the same number of carbon atoms as the parent fatty acid. However, the side chain lacks four carbon atoms than the parent fatty acid that are involved in the formation of cyclobutane ring [Crews et al., 2012]. The radiation-induced 2-ACBs could be estimated through the fatty acid composition of the lipid. The palmitic (C_{16}) and stearic (C_{18}) acids form the 2-dodecylcyclobutanone (2-DCB) and 2-tetradecylcyclobutanone (2-TCB), respectively, upon irradiation. The monounsaturated oleic acid (C_{18}) forms the 2-tetradecenylcyclobutanone (2-TDCB) [Delincée et al., 2002], where the 2-ACB side chain has the same C=C unsaturation as the parent fatty acid [Crews et al., 2012]. Various scientists reported the potential use of radiation-induced 2-ACBs to identify irradiated foods containing fat, such as meat, fish, shrimp, cheese, and egg products [Morehouse et al., 1991; Morehouse and Ku, 1992; Bergaentzle et al., 1994a; Schulzki et al., 1995; Villavicencio et al., 1997]. Since cooking, freezing, decomposition, or oxidation processes do not induce similar chemical products in lipid-containing foods, the 2-ACBs are radiation-specific and may be used as potential radiation-detection markers for the fat-containing foods [Stevenson et al., 1990]. These compounds could also be used as detection-markers in complex food products and food materials of plant origin [Crews et al., 2012].

Considering the potential of 2-ACBs as radiation-detection markers in fat-containing foods, the EN 1785 [2003] standard was developed where the detection method was successfully validated for irradiated chicken, egg, pork, Camembert, and Salmon samples.

Variyar et al. [2008] challenged the radiation-specificity of 2-ACBs because their presence was detected in non-irradiated commercial nutmeg and cashew nut samples. Recently, Chen et al. [2012] conducted a detailed study to verify these findings and reported a contradicting result. They found that the dose-dependent formation of 2-ACBs occurred in irradiated nutmeg samples only. Moreover, the concentrations of 2-ACBs decreased by up to 65–75% during 30 weeks of storage, but the detection was possible in 5 kGy-irradiated samples. In this regard, more extensive research is needed to investigate the positive results (2-ACBs in non-irradiated samples) obtained by Variyar et al. [2008].

Other Radiation-Induced Chemical Changes

Irradiation of different food materials could produce or enhance the concentrations of different gases including carbon monoxide, hydrogen peroxide, hydrogen sulfide, ammonia, and off-odor volatiles [Nam and Ahn, 2003].

The increased concentrations of carbon monoxide in irradiated frozen-chicken meat, beef, and pork were examined by Furuta et al. [1992] using microwave heating to release trapped gas in the headspace and analyzing through gas chromatography. Irradiation can split the water molecule forming hydrogen atoms that can react with O_2 present in the food matrix, thereby producing hydroperoxy radicals. These hydroperoxy radicals can lead to the formation of hydrogen peroxide [Thakur and Singh, 1994].

Off-odors and off-flavors could result from the irradiation of fresh meat depending on the applied-irradiation dose, type of meat, temperature during treatment, oxygen exposure during and/or after the irradiation process, type of packaging, and presence of antioxidative substances [Brewer, 2009]. Unpleasant "metallic", "sulfide", "wet dog", "wet grains" or "burnt" odor in irradiated meat was reported by Huber et al. [1953].

Isooctane-soluble carbonyl compounds in the lipid part and low molecular weight, acid-soluble carbonyls in the protein part of meat could be formed upon irradiation, with the applied dose having a significant influence. However, cooking can decrease the concentrations of these compounds. The 1-heptene and 1-nonene are the most important volatile components showing pronounced effects of irradiation [Brewer, 2009].

Radiation-induced off-flavors in different meat products are ascribed mainly to enhanced lipid oxidation [Trindade et al., 2010; Ahn et al., 2001; Patterson and Stevenson, 1995] However, the combined effect of radiolytic sulfur-volatiles and lipid-oxidation products is important to define the off-flavor in irradiated foods [Jo and Ahn, 2000].

The sulfur-volatiles, such as dimethyl sulfide, dimethyl disulfide, and dimethyl trisulfide, have more influence in the production of off-odors than aldehydes from lipid oxidation [Ahn et al., 2000]. Sulfur-volatiles are difficult to detect in aerobically-packaged than vacuum-packaged meats as they are highly volatile and usually decrease with time after the treatment [Ahn, et al., 2000]. Dimethyltrisulfide is a major off-odor compound, producing fishy and putrid odors, followed by bismethylthiomethane (sulfurous) in irradiated meat products.

The formation and dispersion of free radicals are reduced at low temperature during irradiation, resulting in a favorable effect on the production of odor/flavor compounds. In addition, the detrimental effects of irradiation could be greatly reduced by the removal of oxygen (vacuum packaging), use of inert gases (nitrogen) in packaging, application of antioxidants, and favorable post-irradiation storage conditions including re-packaging or double packaging in oxygen permeable film [Brewer, 2009].

Aromatic amino acids, such as phenylalanine and tyrosine, are present in various food proteins. The reactions of these aromatic amino acids with hydroxyl radicals (from radiolysis of water molecule) form ortho- and meta-tyrosine that are not naturally present [Chauhan et al., 2009]. Karam and Simic [1988] first suggested the use of *o*-tyrosine to detect irradiated meats and shellfish. Meier et al. [1990] reported a simple and fast way to identify irradiated chicken meat through *o*-tyrosine, where the formation of *o*-tyrosine was related to the applied dose, dose rate, and temperature during irradiation.

However, non-irradiated chicken samples have also shown low amounts of *o*-tyrosine. Miyahara and others [2002] also used this technique for irradiated bone-in chicken and reported a good correlation with the ESR results. However, the occurrence of *o*-tyrosine in non-irradiated food samples in low quantities limits its application as a screening marker and requires a confirmatory approach (ESR or TL analysis) for a more reliable result [Meyer et al. 1990].

BIOLOGICAL METHODS

DNA Comet Assay

Ionizing radiations can affect the integrity of the cell structures and particularly target the cellular DNA. Three kinds of radiation-induced changes in DNA are usually observed: (i) double-strand breaks, (ii) single-strand breaks, and (iii) base damage, which can result in the inactivation of living cells, inhibition of growth, and other drastic effects due to damage of molecules that regulate important cell processes [von Sonntag 1987]. The major cause of these changes is the radiation-induced hydroxyl radical produced from the splitting of the water molecule. Considering the potential of radiation-induced DNA damage in food materials, different detection techniques were devised. One of the effective techniques in determining the radiation-induced damage is microgel electrophoresis [Stewart, 2010]. A single cell or nuclei is extracted from food materials and embedded in agarose gel, enabling the analysis of DNA leakage. A compact nuclei with no or slight tails are usually observed in cells from non-irradiated samples (Figure 15). However, the radiation-induced damage causes fragmentation of DNA molecule that will leak out from the nuclei during electrophoresis and form a tail in the direction of the anode depending on the applied irradiation doses. Different foods of animal origin, such as poultry meat, beef, pork, chevon, and mutton, and of plant origin, such as almond, fig, lentil, soy bean, strawberries, grapefruit, and linseed, were successfully analyzed using this technique [Cerda et al., 1993; 1997; Delincee, 1996; 1998]. Cerda [1998] reported the identification of irradiated frozen meat using the DNA comet assay, where the irradiated cells showed comets with long tails, while the non-irradiated cells revealed no or very short tails (Figure 15).

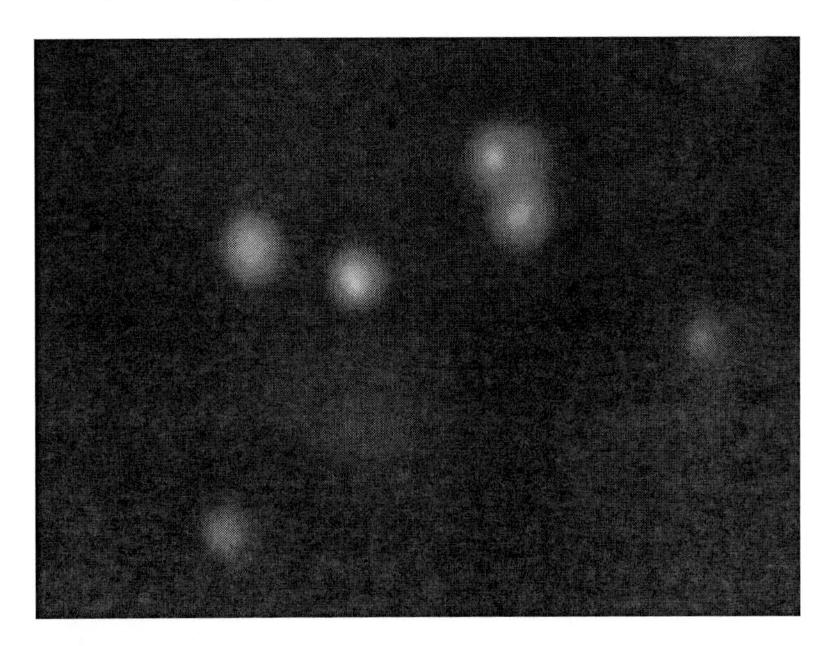

Figure 15. (Continued).

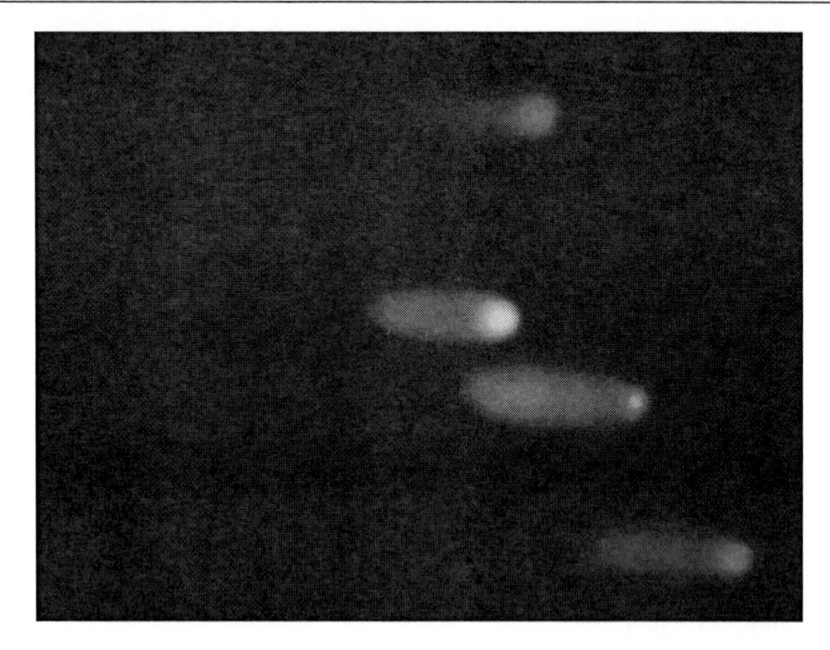

Figure 15. DNA comet assay results from non-irradiated (above; round and intact) and irradiated samples (below; with clear tails).

Similar findings were also reported by Chung et al. [2002; 2004b] in e-beam irradiated (0.5-2 kGy) perilla seeds and gamma-irradiated (1-10 kGy) ostrich meat. Khan and co-workers [2002] reported the successful identification of irradiated spices using DNA comet assay. The irradiated spices exhibited radiation-damaged DNA that stretched or migrated out of the cells as a comet, whereas the non-irradiated spices showed round or conical spots after electrophoresis.

Jo and Kwon [2006] used the same technique to characterize gamma-irradiated (0-2 kGy) kiwifruits where more clear differences were found in the seed samples. The shape, length, and intensity of the DNA tails were dependent of the applied doses. Similar results were also observed in irradiated onions [Alvarez et al., 2007].

Verbeek et al. [2007] used automated image-analyzing system to determine the DNA damage. This system gave a dose-dependent discrimination of irradiated and non-irradiated foods and showed the possibility of dose estimation using this technique. This identification method was successfully validated as a screening approach for irradiated (1-5 kGy) food of animal origin, such as chicken bone marrow, chicken, and pork samples, and for irradiated (0.2-5 kGy) food of plant origin (almonds, figs, lentils, linseed, rose pepper, sesame seeds, soybeans, and sunflower seeds), and standardized as EN 13784 [2002]. Cerda et al. [1997] reported that the lysis of the cell wall was not achieved for the mushroom spores of the *Agaricus bisporus*, and thus the comet assay could not be useful in this case.

The application of this method is limited to food materials that have not been treated with heat or other processing, which could cause DNA damage, resulting in comets similar to those from irradiated samples. In this regard, DNA comet assay could only be used as a screening approach and positive results must be confirmed by some reliable (TL and ESR analyses) techniques [13784, 2002; Chung et al., 2004a].

Direct Epifluorescent Filter Technique Combined with Aerobic Plate Count (DEFT/APC)

Ionizing radiations kill microbes by targeting their DNA, resulting in improved hygienic quality and enhanced shelf-life of food materials [Akram and Kwon, 2010]. Therefore, it is possible to develop a screening technique using DEFT/APC, providing an easy discrimination of food materials which underwent a processing method that resulted in microbial decontamination.

The APC count, which shows the number of viable microorganisms, and DEFT count, the total number of viable and non-viable microbial cells, from the sample are determined and compared [EN 13783, 2001]. The microorganisms are concentrated on a membrane filter and stained with a fluorochrome, acridine orange (AO). A blue light at 450 to 490 nm is used to obtain an orange and orange-yellow fluorescence from the stained-microbial cells. The total number of viable and non-viable microbial cells (DEFT count) is determined using an epifluorescence microscope, while another portion of the same sample is analyzed for APC [EN 13783, 2002].

For irradiated samples, the difference between the DEFT and APC counts is normally ≥ 3-4 log units due to the death of microorganisms upon irradiation. However, similar differences could also be observed when other techniques, such as heat or chemical treatments, were used since these methods also cause microbial reduction [Stewart, 2010]. Hence, the DEFT/APC method has limited applications and could be used as a screening approach only, in which the results must be confirmed using a more reliable detection technique.

In the case of non-irradiated sample, comparable results from both analyses (DEFT and APC) were found. However, the significant difference (very low APC) indicated an application of a decontamination treatment [Jones et al., 1994]. Fumigation or heat treatment could provide the same results. Also, this technique is difficult to apply in samples with very low initial microbial load (APC $< 10^3$ CFU/g).

Microbial inhibition due to some natural antimicrobial components in spices, such as cloves, cinnamon, garlic, and mustard seeds, also restricts the application of this technique [Stewart, 2010]. An EN protocol [EN 13783, 2002] was standardized after the successful validation (through interlaboratory trials) of this method as a screening approach for different irradiated (5 and 10 kGy) herbs and spices (allspice, whole and powdered black pepper, whole white pepper, paprika powder, cut basil, cut marjoram, and crushed cardamom).

Limulus Amebocyte Lysate Test Combined with Gram-Negative Bacterial Count (LAL/GNB)

This technique particularly targets the Gram-negative bacterial population in the sample. First, the viable count of Gram-negative bacteria in the test sample is determined. In the second step, the concentration of bacterial endotoxin from the surfaces of Gram-negative bacteria as lipopolysaccharides is determined using LAL test.

The concentration of lipopolysaccharides represents the viable and dead Gram-negative bacteria. In the case of irradiated sample, the difference between Gram-negative bacteria and endotoxin is very prominent, showing the application of an antimicrobial technique. However, only a screening approach is possible as the death of microorganisms could be due to reasons other than irradiation. The production of endotoxin by Gram-negative bacteria before their death could also create a problem for an accurate estimation. Scotter and co-workers [1994] applied this technique in chicken meat and suggested a successful discrimination approach on the basis of microbiological profiles. However high-quality meat with very low initial microbial load might be difficult to discriminate from irradiated samples using this method. This technique could provide only an indication of irradiation treatment where positive results need a reliable confirmation.

This approach has been validated through interlaboratory trials for irradiated boneless-chicken breasts with skin and boneless-chicken fillets and standardized as an EN protocol [EN 14569, 2001]. Although the applicability of this method is limited and can only be applied as a screening method (requiring a confirmation approach), it is still useful to investigate the microbial profile of a sample before and after irradiation [Stewart, 2010].

Other Potential Methods to Develop a Screening Approach for Irradiated Foods

The viscosity change as a detection marker in irradiated (8 kGy) spices was first reported by Mohr and Wichmann [1985]. The radiation-induced physical degradation of food macromolecules resulting in reduced viscosity of solution was also observed by various scientists [Glidewell et al., 1993; Polonia et al., 1995; Schreiber et al., 1993; Stevenson and Stewart, 1995]. It could be used as an easy, cheap and time-efficient method to detect radiation treatment [Schreiber et al., 1993]. The method was found effective for the spices with considerable starch content, where results could also be affected by shear rate, temperature, pH, and type of viscometer. However, the change in viscosity is not radiation-specific phenomenon and may be induced by other routine food preservation methods particularly those involving heat treatment. Thus, the results should be confirmed using ESR or TL analysis [Molins, 2001]. Barabassy and co-workers [1996] reported that an irradiation dose of 8 kGy showed a clear reduction in the solution viscosity of black and white pepper, ginger, and nutmeg, with the best results observed in irradiated pepper samples [Hayashi, 1996]. Using this method, the irradiation history of dried vegetable starches, fish, and other seafood samples could also be screened [Farkas et al., 1990]. The significant effect of high pH (13.8) for the improved discrimination of irradiated and non-irradiated samples was reported by Hayashi and Todoriki [1996].

Kawamura et al. [1996] presented valuable results through a collaborative study using the half-embryo test and accounted that irradiated and non-irradiated citrus fruits can be discriminated using an easy and low-cost half-embryo test. Twelve participating laboratories determined the effect of irradiation (0, 0.2, and 0.5 kGy) on the citrus seeds through half-embryo test. Results showed that 92% and 98% of the samples after 4 and 7 days of incubation, respectively, were correctly identified based on their irradiation history. This test was simple, economical and environment friendly. In addition, the test was effective for low dose (0.15 kGy) irradiation without the involvement of sophisticated equipment.

Selvan and Thomas [1999] reported the effect of irradiation on the root length and root number in onion bulbs. The irradiated bulbs, grown on 0.2% agar, showed a drastic inhibition in root length only. The developed method was easy and cost effective, which was mainly based on the root numbers and root length to identify irradiated onions and shallots. The chemical sprout inhibitor (maleic hydrazide) was also used to compare the results. Unlike the irradiated samples, the maleic hydrazide-treated samples did not exhibit root inhibition.

Considering the effect of irradiation on the volatile compound profile, different scientists have shown the applicability of E-nose analysis to identify irradiated tomatoes [Winquist et al., 1995], anchovy sauce [Kim et al., 2004], red pepper powder [Lee et al., 2005], and ground pork [Kim et al., 2008]. Recently, Akram et al. [2012c] presented valuable findings using the principal component analysis of E-nose data to characterize the king oyster mushroom samples in accordance to their irradiation history, where the responses were dependent on irradiation doses.

Stefanova et al. [2011] reported the dose-dependent effect of irradiation on the fatty acid profile as characterized by ^1H NMR spectroscopy. The results showed a dose-dependent increase in saturated fatty acids and a decrease in polyunsaturated fatty acids in the triacylglycerol composition of irradiated meat samples. The decreasing trend in polyunsaturated fatty acyl groups was attributed to the decrease in the oxidative stability of the meat fat upon irradiation. Zoumpoulakis et al. [2012] also observed the effect of irradiation on the triglyceride composition in white sesame seeds and they suggested that the 2D DOSY NMR technique could be applied as a qualitative detection tool for low-dose irradiated samples.

Practical Limitations and Research Trends for the Improved Methods

All of the available identification methods have some limitations in terms of their application range and effectiveness during different post-irradiation conditions and thus none has the potential to be applied to all kinds of foods [Delincée, 1998]. Luminescence techniques depend on the contaminating minerals which are usually not inherent to food itself. Moreover, luminescence characteristics are not stable during severe processing and storage conditions (temperature, light, and time) [Kwon et al., 2011c]. ESR spectroscopy can only be applied to solid foods containing cellulose, sugar, and bone. In addition, ESR spectra are also not stable during the post-irradiation storage. The ESR results may be complicated due to the presence of natural ESR signals (e.g. Mn^{2+}) [Ahn et al., 2012f]. Radiation-induced chemical markers, such as hydrocarbons and 2-ACBs, can be detected only in fat-containing food materials. Hence, the applicability and reliability of each method under different circumstances are a crucial aspect to be investigated [Arvanitoyannis, 2010]. Research is also needed for the identification of different irradiated ingredients in a complex-formula foods and stability of radiation-induced markers during post-irradiation processing and storage [Ahn et al., 2012b, 2012c]. Dose-estimation is also an interesting aspect of research as there is a variation in approved doses for the same food product in different countries [D'Oca et al., 2009]. A worldwide database for the validation and selection of specific detection techniques for a particular food item is also needed [Kwon et al., 2010].

REFERENCES

Ahn, D. U., Jo, C., Du, M., Olson, D. G., and Nam, K. C. (2000). Quality characteristics of pork patties irradiated and stored in different packaging and storage conditions. *Meat Science*, 56, 203–209.

Ahn, D. U., Nam, K. C., Du, M., and Jo, C. (2001). Volatile production in irradiated normal, pale soft exudative (PSE) and dark firm dry (DFD) pork under different packaging and storage conditions. *Meat Science*, 57, 419–426.

Ahn J. J., Kim, G. R., Jin, Q. W. and Kwon, J. H. (2009a). Analytical properties of electron spin resonance after irradiation of seasonings with different radiation sources. *Korean Journal of Food Preservation*, 16, 385–391.

Ahn, J. J. Lee, J. W., Chung, H. W., and Kwon, J. H. (2009b). Analytical Characteristics of electron spin resonance for identifying irradiated Ramen soup with radiation sources. *Korean Journal of Food Science and Technology*, 41, 131–135.

Ahn, J. J., Kim, G. R., Akram, K., Kim, K. S., and Kwon, J. H. (2012a). Effect of storage conditions on photostimulated luminescence of irradiated garlic and potato. *Food Research International*, 47, 315-320.

Ahn, J. J., Kim, G. R., Akram, K., Kim, K. S., and Kwon, J. H. (2012b). Luminescence characteristics of minerals separated from irradiated onions during storage under different light conditions. *Radiation Physics and Chemistry*, doi: org/10.1016/j. radphyschem. 2012.02.002.

Ahn, J. J., Akram, K., Lee, J., Kim, K. S., and Kwon, J. H. (2012c). Identification of a gamma-irradiated ingredient in Korean barbecue sauce by thermoluminescence analysis before and after pasteurization. *Journal of Food Science*, 77, 476-480.

Ahn, J. J., Akram, K., Jeong, M. S., Kwak, J. Y., and Kwon, J. H. (2012d). Radiation-induced thermoluminescence characteristics of feldspar following different heat and microwave treatments. *Journal of Luminescence*, 132, 1964–1968.

Ahn, J. J., Akram, K., and Kwon, J. H. (2012e). Electron spin resonance analyses of grinding- and radiation-induced signals in raw and refined sugars. *Food Analytical Methods*, doi: 10.1007/s12161-012-9364-z.

Ahn, J. J., Akram, K., Kim H. K., and Kwon J. H. (2012f). Electron spin resonance spectroscopy for the identification of irradiated food with complex ESR signals. *Food Analytical Methods*, doi: 10.1007/s12161-012-9440-4.

Akram, K., Ahn, J. J., and Kwon, J. H. (2012a). Identification and characterization of gamma-irradiated dried *Lentinus edodes* using ESR, SEM, and FTIR analyses. *Journal of Food Science*, doi: 10.1111/j.1750-3841.2012.02740.x.

Akram, K., Ahn, J. J., Kim, G. R., and Kwon, J. H. (2012b). Applicability of different analytical methods for the identification of gamma-irradiated fresh mushrooms during storage. *Food Science and Biotechnology*, 21, 573-679.

Akram, K., Ahn, J. J., Yoon, S. R., Kim, G. R., and Kwon, J. H. (2012c). Quality attributes of *Pleurotus eryngii* following gamma irradiation. *Postharvest Biology and Technology*, 66, 42–47.

Akram, K., and Kwon, J. H. (2010). Food irradiation for mushrooms: A review. *Journal of Korean Society of Applied Biological Chemistry*, 53, 257–265.

Aleksieva, K., Georgieva, L., Tzvetkova, E., and Yordanov, N. D. (2009). EPR study on tomatoes before and after gamma-irradiation. *Radiation Physics and Chemistry,* 78, 823–825.

Alvarez, D. L. M., Miranda, E. F. P., Palacio, S. C., and Enriquez, I. I. (2007). Detection of irradiated onion by means of the comet assay. *International Nuclear Atlantic Conference–INAC* 2007 Santos, SP, Brazil, ISBN: 978-85-99141-02-1.

Arvanitoyannis, I. S. (2010). Irradiation of food commodities: Techniques, ppplications, Detection, legislation, safety and consumer Opinion). 1st Edition. London: *Academic Press an imprint of Elsevier.*

Barabassy, S, Sharif, M, Farkas, J, Konez, A, Formanek, Z, and Kaffka, K. (1996). Attempts to elaborate detection methods for some irradiated food and dry ingredients. In: McMmurry, C. H., Stewart, E. M., Gray, R., and Pearce, J. (Eds.), Detection methods for irradiated foods-current status (Special Publication 171). pp. 185–201. Cambridge: *Royal Society of Chemistry.*

Beneitez, P., Correcher, V., Millán, A., and Calderón, T. (1994). Thermoluminescence analysis for testing the irradiation of spices. *Journal of Radioanalytical and Nuclear Chemistry*, 185, 401–410.

Bergaentzle, M., Hasselmann, C., and Marchioni, E. (1994). Detection of irradiated foods by the mitochondrial DNA method. *Food Science and Technology Today*, 8, 111–113.

Bhatti, I. A., Akram, K., and Kwon, J. H. (2012). An investigation into gamma-ray treatment of shellfish using electron paramagnetic resonance spectroscopy. *Journal of the Science of Food and Agriculture*, 92, 759–763.

Bortolin, E., Boniglia, C., Calicchia, A., Alberti, A., Fuochi, P., and Onori, S. (2007). Irradiated herbs and spices detection: light-induced fading of the photo- stimulated luminescence response. *International Journal of Food Science and Technology*, 42, 330–335.

Brewer, M. S. (2009) Irradiation effects on meat flavor-a review. *Meat Science*, 81, 1-14.

Calderón, T., Correcher, V., Millán, A., Beneitez, P., Rendell, H. M., Larsson, M. Townsend, P. D., and Wood, R. A. (1995). New data on thermoluminescence of inorganic dust from herbs and spices. *Journal of Physics: D-Applied Physics*, 28, 415–423.

Calucci, L., Pinzino, C., Zandomeneghi, M., Capocchi, A., Ghiringhelli, S., Saviozzi, F., Tozzi, S., and Galleschi, L. (2003). Effects of γ-irradiation on the free radical and antioxidant contents in nine aromatic herbs and spices. *Journal of Agricultural and Food Chemistry*, 51, 927–934.

Cerda, H. (1998). Detection of irradiated fresh chicken, pork and fish using the DNA comet assay. *Lebensmittel-Wissenschaft and Technologie*, 31, 89–92.

Cerda, H., Delincée, H., Haine, H., and Rupp, H. (1997). The DNA "Comet Assay" as a rapid screening technique to control irradiated food. *Mutation Research*, 375, 167–181.

Cerda, H., von Hofsten, B., and Johanson, K. J. (1993). Identification of irradiated food by microelectrophoresis of DNA from single cells. In: Leonardi, M., Belliardo, J. J., and Raffi, J. J. (Eds.). Recent advances on detection of irradiated food. pp. 401–405. Luxembourg: *Commission of the European Communities.*

Cerv, P., Schara, M., and Ravnik. C. (1972). Electron paramagnetic study of irradiated tooth enamel. *Radiation Research*, 51, 581–589.

Champagne, J. R., and Nawar, W. W. (1969). The volatile components of irradiated beef and pork fats, *Journal of Food Science*, 34, 335–339.

Chauhan, S. K., Kumar, R., Nadanasabapathy, S., and Bawa, A. S. (2009). Detection methods for irradiated foods. *Comprhensive Reviews in Food Science and Food Safety,* 8, 4–16.

Chawla, S. P., and Thomas, P. (2004). Identification of irradiated chicken meat using electron spin resonance spectroscopy. *Journal of Food Science and Technology*, 41, 455–458.

Chen, S., Tsutsumi, T., Takatsuke, S., Matsuda, R., Kameya, H., Nakajima, M., Furuta, M., and Todoriki, S. (2012). Identification of 2-aldylcyclobutanones in nutmeg (*Myristica fragrans*). *Food Chemistry*, 134, 359-365.

Chiaravalle, A. E., Mangiacotti, M., Marchesani G., and Vegliante. G. (2010). Electron spin resonance (ESR) detection of irradiated fish containing bone (gilthead sea bream, cod, and swordfish). *Veterianary Research Commumications*, 34, 149–152.

Chung, H. W., Delincee, H., Han, S. B., Hong, J. H., Kim, H. Y. and Kwon, J. H. (2002). Characteristics of DNA comet, photostimulated luminescence, thermoluminescence and hydrocarbon in perilla seeds exposed to electron beam. *Journal of Food Science*, 67, 2517–2522.

Chung, H. W., Delincée, H., Han, S. B., Hong, J. H., Kim, H. Y., Kim, M. C., Byun, M. W. and Kwon, J. H. (2004a). Trials to identify irradiated chestnut with different analytical techniques. *Radiation Physics and Chemistry*, 71, 181–184.

Chung, H. W., Hong, J. H., Kim, M. C., Marshall, M. R., Jeong, Y., and Han, S. B. (2004b). Detection properties of irradiated ostrich meat by DNA comet assay and radiation-induced hydrocarbons. *Journal of Food Science*, 69, C399–403.

Crews C., Driffield, M., and Thomas. C. (2012). Analysis of 2-alkylcyclobutanones for detection of food irradiation: Current status, needs and prospects. *Journal of Food Composition and Analysis*, doi: 10.1016/j.jfca.2011.11.006.

D'Oca, M. C., Bartolotta, A., Cammilleri, C., Giuffrida, S., Parlato, A., and Di Stefano, V. (2009). The additive dose method for dose estimation in irradiated oregano by thermoluminescence technique. Food Control, 20, 304–306.

D'Oca, M. C., Bartolotta, A., Cammilleri, C., Giuffrida, S., Parlato, A., and Di Stefano, V. (2010). A practical and transferable methodology for dose estimation in irradiated spices, based on thermoluminescence dosimetry. *Applied Radiation and Isotopes*, 68, 639–642.

D'Oca, M. C., Bartolotta, A., Cammilleri, M. C., Brai, M., Marrale, M., Triolo, A., and Parlato, A. (2007). Qualitative and quantitative thermoluminescence analysis on irradiated oregano. *Food Control*, 18, 996–1001.

de Jesus, E. F. O., Rossi, A. M. and Lopes, R. T. (1996). Influence of sample treatment on ESR signal of irradiated citrus. *Applied Radiation Isotopes* 47, 1647-1653.

de Jesus, E. F. O., Rossi, A. M., and Lopes, R. T. (1999). An ESR study on identification of gamma-irradiated kiwi, papaya and tomato using fruit pulp. *International Journal of Food Science and Technology*, 34, 173–178.

de Jesus, E. F. O., Rossi, A. M., and Lopes, R. T. (2000). Identification and dose determination using ESR measurements in the flesh of irradiated vegetal products. *Applied Radiation Isotopes*, 52, 1375–1383.

Deighton, N., Glidwell, S. M., Goodman, B. A., and Morrison, I. M. (1993). Electron paramagnetic resonance of gamma-irradiated cellulose and lignocellulosic material. *International Journal of Food Science and Technology*, 28, 45–55.

Delincée H. (1996). Introduction to DNA methods for identification of irradiated foods. In: McMurray CH, Stewart, EM, Gray R, and Pearce J. (Eds.). Detection Methods for Irradiated Foods - Current Status. pp. 345–348. Cambridge: *Royal Society of Chemistry.*

Delincée, H. (1998). Detection of irradiated food DNA fragmentation in grapefruits. *Radiation Physics and Chemistry*, 52, 135–139.

Delincée, H. (2002). Analytical methods to identify irradiated food: A review. *Radiation Physics and Chemistry*, 63, 455–458.

Delincée, H., and Soika. C. (2002). Improvement of the ESR detection of irradiated food containing cellulose employing a simple extraction method. *Radiation Physics and Chemistry*, 63, 737–741.

Desrosiers, M. F. (1989). Gamma-irradiated seafoods: Identification and dosimetry by electron paramagnetic resonance spectroscopy. *Journal of Agricultural and Food Chemistry*, 37, 96–100.

Desrosiers, M. F. (1996). Current status of the EPR method to detect irradiated food. *Applied Radiation and Isotopes*, 47, 1621–1628.

Diehl, J. F. (1990) Lebensmittelbestrahlung – Risiko oder Fortschritt? Industrielle Obst– und Gemüseverwertung, 75, 58–63.

Diehl, J. F. (1995). *Safety of Irradiated Foods*. 2nd Edition. New York: CRC Press.

Dodd, N. J. F., Swallow, A. J., and Ley. F. J. (1985). Use of ESR to identify irradiated foods. *Radiation Physics and Chemistry*, 26, 451–453.

Duller, G. A. T., Bøtter-Jensen, L., Kohsiek, P., and Murray A. S. (1999). A high-sensitivity optically stimulated luminescence scanning system for measurement of single sand-sized grain. *Radiation Protection Dosimetry*, 84, 325–330.

Ehlermann, D. A. E. (1999) Eröffnung und Begrüßung. Lebensmittelbestrahlung – 5. Deutsche Tagung. Berichte der Bundesforschungsanstalt für Ernährung. BFE–R—99–01. *Bundesforschungsanstalt für Ernährung*, Karlsruhe (Deutschland). 1–4.

Elahi, S., Straub, I., Thurlow, K., Farnell, P., and Walker, M. (2008). Referee analysis of suspected irradiated food. *Food Control*. 19, 269–277.

Empis, J. M. A., Silva, H. A., Nunes, M. L., and Andrade, E. M. (1995). Detection of irradiated fish using EPR of fish bone signal intensity and stability. *Fish Research*, 21, 417–475.

EN13708. (2001). Foodstuffs-Detection of irradiated food containing crystalline sugar by ESR spectroscopy. Brussels, Belgium: *European Committee of Standardization* (CEN).

EN13751. (2009). Foodstuffs-Detection of irradiated food using photostimulated luminescence. Brussels, Belgium: *European Committee of Standardization (CEN).*

EN13783. (2001). Foodstuffs-Detection of irradiated food using direct epifluorescent filter technique/aerobic plate count (DEFT/APC)-screening method. Brussels, Belgium: *European Committee of Standardization (CEN).*

EN13874. (2001). Foodstuffs-DNA coment assay for the detection of irradiated foodstuffs-screening method. Brussels, Belgium: *European Committee of Standardization (CEN).*

EN14569. (2004). Foodstuffs-Microbiological screening for irradiated food using LAL/GNB procedures. Brussels, Belgium: *European Committee of Standardization (CEN).*

EN1784. (2003). Foodstuffs-Detection of irradiated food containing fat-Gas chromatographic analysis of hydrocarbons. Brussels, Belgium: *European Committee of Standardization (CEN).*

EN1785. (2003). Foodstuffs-Detection of irradiated food containing fat-Gas chromate-graphic /mass spectrometric analysis of 2-alkylcyclobutanones. Brussels, Belgium: *European Committee of Standardization (CEN).*

EN1787. (2000). Foodstuffs-Detection of irradiated food containing cellulose by ESR spectroscopy. Brussels, Belgium: *European Committee of Standardization (CEN).*

EN1788. (2001). Foodstuffs-Thermoluminescence detection of irradiated food from which silicate minerals can be isolated. Brussels, Belgium: *European Committee of Standardization (CEN).*

Fan, X., and Sommers, C. H. (2006). Effect of gamma radiation on furan formation in ready-to eat products and their ingredients. *Journal of Food Science,* 71, C407–C412.

FAO (1965). The Technical Basis for Legislation on Irradiated Food. Report of a Joint FAO/IAEA/WHO Expert Committee; pp. 56. Rome, 21–28 April 1964. *FAO Atomic Energy Series No. 6*, FAO, Rome (Italy).

Farkas, J., and Moh´acsi-Farkas, C. (2011). History and future of food irradiation. *Trends in Science and Technology*, 22, 121–126.

Farkas, J., Koncz, A., and Sharif, M. M. (1990). Identification of irradiated dry ingredients on the basis of starch damage. *Radiation Physics and Chemistry*, 35, 324–328.

Fuochi P. G., Alberti, A., Bortolin, E., Corda, U., La Civita, S., and Onori, S. (2008). PSL study of irradiated food: NaCl as possible reference material. *Radiation Measurements*, 43, 483–486.

Furuta, M., Dohmaru, T., Katayama, T., Toratani, H., and Takeda. A. (1992). Detection of irradiated frozen meat and poultry using carbon monoxide gas as a probe. *Journal of Agriculture and Food Chemistry*, 40, 1099–1100.

Garcia-Guinea, J., Finch, A., Can, N., Hole, D., and Townsend, P. (2007). Orientation dependence of the ion beam and cathodoluminescence of albite. *Physica Status Solidi*, 4, 910–913.

Gastelum, S., Osuna, I., Melendrez, R., Cruz-Zaragoza, E., Chernov, V., Calderon T., and Barboza-Flores, M. (2002). Application of a thermoluminescence method for the detection of irradiated spices. *Radiation Protection Dosimetry*, 101, 137–140.

Glidewell, S. M., Deighton, N., Goodman, B. A., and Hillman, J. R. (1993). Detection of irradiated food: A review. *Journal of the Science of Food and Agriculture*, 61, 281–300.

Glimcher, M. J. (1984). Recent studies of the mineral phase in bone and its possible linkage in organic matrix by protein-bound phosphate bands. *Philosophical Transactions of the Royal Society B: Biological Sciences,* 304, 479–508.

Goodman, B. A., McPhail, D. B., and Duthie. D. M. L. (1989). Electron spin resonance spectroscopy of some irradiated foodstuffs. *Journal of the Science of Food and Agriculture*, 47, 101–111.

Gordy, W., Ard, W. B., and Shields, H. (1955). Microwave spectroscopy of biological substances. I. paramagnetic resonance in X-irradiated amino acids and proteins. *The Proceedings of the National Academy of Sciences USA*, 41, 983–996.

Gray, R. and Stevenson, M. H. (1989). Detection of irradiated deboned turkey meat using electronspinresonance spectroscopy. *Radiation Physics and Chemistry*, 34, 899–902.

Gray, R., and Stevenson, M. H. (1990). Effect of length and temperature of storage on ESR signal from various bones in irradiated chicken carcasses. *International Journal of Food Science and Technology*, 25, 506–511.

Hayashi T. (1996). Applicability of viscosity measurement to the detection of irradiated peppers. In: McMurray CH, Stewart EM, Gray R, and Pearce J, (Eds.), Detection methods for irradiated foods–current status (Special publication 171). pp. 215–228. Cambridge: *Royal Society of Chemistry.*

Hayashi, T., and Todoriki, S. (1996). Detection of irradiated peppers by viscosity measurement at extremely high pH. *Radiation Physics and Chemistry*, 48, 101–104.

Huber, W., Brasch, A., and Waly, A. (1953). Effect of processing conditions on organoleptic changes in foodstuffs sterilized with high intensity electrons. *Food Technology*, 7, 109–115.

Hwang, K. T. (1999). Hydrocarbons detected in irradiated pork, bacon and ham. *Food Research International*, 32, 389–394.

IAEA (1966). Food Irradiation. *Proceedings of the International Symposium on Food Irradiation jointly organized by the International Atomic Energy Agency and the Food and Agriculture Organization of the United Nations and held in Karlsruhe*, 6–10 June 1966. pp. 956. IAEA, Vienna (Austria).

Jo, C., and Ahn, D. U. (2000). Production of volatiles from irradiated oil emulsion systems prepared with amino acids and lipids. *Journal of Food Science*, 65, 612–616.

Jo, D., and Kwon, J. H. (2006). Detection of radiation-induced markers from parts of irradiated kiwifruits. *Food Control*, 17, 617–621.

Jones, K., MacPhee, S., Turner, A., Stuckey, T., and Betts, R. (1994). The direct epifluorescence filter technique (DEFT)/ aerobic plate count (APC): a screening method for the detection of irradiated frozen stored foods. A collaborative trial. *Food Science and Technology Today*, 9, 141–144.

Josephson, E. S., and Peterson, M. S. (2000). *Preservation of food by ionizing radiation (I)*. Boca Raton, Florida: CRC Press.

Karakirova, Y., Yordanov, N. D., De Cooman, H., Vrielinck, H., and Callens. F. (2010). Dosimetric characteristics of different types of saccharides: an EPR and UV spectrometric study. *Radiation Physics and Chemistry*, 79, 654–659.

Karam, L. R., and Simic, M. G. (1988). Detection of irradiated meats: A use of hydroxyl radical biomarkers. *Analytical Chemistry*, 60, 1117A–1119A.

Kawamura, Y., Sugita, T., Yamada, T., and Saito Y. (1996). Half-Embryo test for identification of irradiated citrus fruit: Collaborative study. *Radiation Physics and Chemistry*, 49, 665–668.

Khan, A. A., Khan, H. M., and Delincée, H. (2002). Detection of radiation treatment of beans using DNA Comet assay. *Radiation Physics and Chemistry*, 63, 407–410.

Kikuchi, M., Hussain, M. S., Morishita, N., Ukai, M., Kobayashi, Y., and Shimoyama, Y. (2010). ESR study of free radicals in mango. Spectrochimica Acta Part A: *Molecular and Biomolecular Spectroscopy*, 75, 310–313.

Kikuchi, M., Shimoyama, Y., Ukai, M., and Kobayashi, Y. (2011). ESR detection procedure of irradiated papaya containing high water content. *Radiation Physics and Chemistry*, 80, 664–667.

Kim, B. K., Akram, K., Kim, C. T., Kang, N. R., Lee, J. W., Ryang, J. H., and Kwon, J. H. (2012). Identification of low amount of irradiated spices (red pepper, garlic, ginger powder) with luminescence analysis. *Radiation Physics and Chemistry*, doi: 10.1016/j.radphyschem.2012.01.023

Kim, H., Cho, W. J., Ahn, J. S., Cho, D. H., and Cha, Y. J. (2005). Identification of radiolytic marker compounds in the irradiated beef extract powder by volatile analysis. *Micro chemical Journal*, 80, 127–137.

Kim, J. H., Ahn, H. J., Yook, H. S., Kim, K. S., Rhee, M. S., Ryu, G. H., and Byun, M. W. (2004). Color, flavor, and sensory characteristics of gamma-irradiated salted and fermented anchovy sauce. *Radiation Physics and Chemistry*, 69,179–187.

Kim, J. H., Lee, J. W., Shon, S. H., Jang, A., Lee, K. T., Lee, M., and Jo, C. (2008). Reduction of volatile compounds and off-odor in irradiated ground pork using a charcoal packaging. *Journal of Muscle Foods*, 19, 194–208.

Kim, M. O., Kwon, J. H., and Bhatti I. A. (2010). Comparison of radiation-induced hydrocarbons for the identification of irradiated perilla and sesame seeds of different origins. *Journal of the Science of Food and Agriculture*, 90, 30–35.

Kume, T., Furuta, M., Todoriki, S., Uenoyama, N., and Kobayashi, Y. (2009). Status of food irradiation in the world. *Radiation Physics and Chemistry*, 78, 222–226.

Kwon, J. H. (2010). Safety and understanding of irradiated food. 1st edition. Seoul (Korea): *Korea Food Safety Research Institute.*

Kwon, J. H., Akram, K., Nam, K. C., Lee, E. J., and Ahn, D. U. (2011a). Evaluation of radiation-induced compounds in irradiated raw or cooked chicken meat during storage. *Poultry Science*, 90, 2578–2583.

Kwon, J. H., Lee, E. J., Kausar, T., and Ahn, D. U. (2011b). Effect of fat substitute and plum extract on radiation-induced hydrocarbons and 2-alkylcyclobutanones in freeze-dried beef patties. *Korean Journal of Food Science and Animal Resources*, 31, 858–864.

Kwon, J. H., Chung, H. W., Kim, B. K., Ahn, J. J., Kim, G. R., Jo, D., and Ahn, K. A. (2011c). Research and application of identification methods for irradiated foods. *Journal of the Korean Society of Food Hygiene and Safety*, 6, 11–27.

Kwon, J. H., Chung, H. W., and Byun, M. W. (2000). ESR spectroscopy for detecting gamma- irradiated dried vegetables and estimating absorbed doses. *Radiation Physics and Chemistry*, 57, 319–324.

Kwon, J. H., Kausar, T., Lee, J., Kim, H. K., and Ahn, D. U. (2007). The microwave-assisted extraction of fats from irradiated meat products for the detection of radiation-induced hydrocarbons. *Food Science and Biotechnology*, 16, 150–153.

Kwon, J. H., Kwon, Y. J., Kausar, T., Nam, K. C., Min, B. R., Lee, E. J., and Ahn, D. U. (2012). Effect of cooking on radiation-induced chemical markers in beef and pork during storage. *Journal of Food Science*, 77, 211–215.

Le Tellier, P. R., and Nawar, W. W. (1972). 2-Alkylcyclobutanones from radiolysis of lipids. *Lipids*, 7, 75–76.

Lee, H. M., Kwak, B. M., Ahn, J. H., and Jeong, S. H. (2010). Development of ESR method for gamma-irradiated lactose powders. *Journal of Food Engineering*, 100, 25–31.

Lee, J. H., Lee, K. T., and Kim, M. R. (2005). Effect of gamma-irradiated red pepper powder on the chemical and volatile characteristics of kakdugi, a Korean traditional fermented radish kimchi. *Journal of Food Science*, 70, C441–C447.

Lee, J., Kausar, T., Kim, B. K., and Kwon, J. H. (2008a). Detection of γ-irradiated sesame seeds before and after roasting by analyzing photostimulated luminescence, thermoluminescence, and electron spin resonance. *Journal of Agricultural and Food Chemistry*, 56, 7187–7188.

Lee, J., Kausar, T., and Kwon, J. H. (2008b). Characteristic hydrocarbons and 2-alkylcyclobutanones for detection γ-irradiated sesame seeds after steaming, roasting, and oil extraction. *Journal of Agricultural and Food Chemistry*, 56, 10391–10395.

Li, B., and Li, S. H. (2008). Investigations of the dose-dependent anomalous fading rate of feldspar from sediments. *Journal of Physics D: Applied Physics*, 41, 1–15.

Loaharanu, P. (2003). Irradiated Foods, Fifth Edition. *American Council on Science and Health*. N. Y. USA.

Malec-Czechowska, G., Strzelczak, A. M., Dancewicz, W., Stachowicz, W., and Delincée. H. (2003). Detection of irradiation treatment in dried mushrooms by photostimulated luminescence, EPR spectroscopy and thermoluminescence measurements. *European Food Research Technology*, 216,157–165.

Marchioni, E. (2006). Detection of irradiated foods. In: Sommers CH, Fan X, editor. *Food Irradiation Research and Technology*. pp. 85–103. Ames, IA: Blackwell Publishing;

Marchioni, E., Horvatovich, P., Ndiaye, B., Miesch, M., and Hasselmann, C. (2002). Detection of low amount of irradiated ingredients in non-irradiated precooked meals. *Radiation Physics and Chemistry*, 63, 447–450.

Maurer, K. F. (1958). Zur Keimfreimachung von Gewürzen. *Ernährungswirtschaft*, 5, 45–47.

McKeever, S. W. S. (1985). *Thermoluminescence of Solids*. Cambridge: Cambridge University Press.

McMurray C. H., Stewart, E. M., Gray, R., and Pearce, J. (1996). Detection methods for irradiation foods-current status. *Royal Society of Chemistry*, Special publication 171. Cambridge, U.K.

Meier, W., Burgin, R., and Frohlich, D. (1990). Analysis of o-tyrosine as a method for the identification of irradiated chicken and the comparison with other methods (Analysis of volatiles and ESR spectroscopy). *Radiation Physics and Chemistry*, 35, 332–336.

Miyahara, M., and Miyahara, M. (2002). Effects of gamma ray and electron-beam irradiations on survival of anaerobic and facultatively anaerobic bacteria. *Kokuritsu Iyakuhin Shokuhin Eisei Kenkyusho Hokoku*, 120, 75–80.

Mohr, F., and Wichmann, G. (1985). Viskositätserniedrigungen als Indiz für eine Cobalt bestrahlung an Gewürzen? *Gordian*, 85, 96.

Molins, R. A. (2001). Food irradiation: Principles and applications. *Hoboken*, NJ: John Wiley and Sons.

Morehouse, K. M., Ku, Y., Abrecht, H. L., and Yang, G. C. (1991). Gas chromatographic and electron spin resonance investigations of gamma irradiated frog legs. *Radiation Physics and Chemistry*, 38, 62–68.

Morehouse, K. M., and Ku, Y. (1992). Gas chromatographic and electron spin resonance investigations of γ-irradiated shrimp. *Journal of Agricultural and Food Chemistry*, 40, 1963–1971.

Nam, K. C., and Ahn, D. U. (2003). Use of antioxidants to reduce lipid oxidation and off-odor volatiles of irradiated pork homogenates and patties. *Meat Science*, 63, 1–8.

Nam, K. C., Lee, E. J., Ahn, D. U., and Kwon, J. H. (2011). Dose-dependent changes of chemical attributes in irradiates in irradiated sausages. *Meat Science*, 88, 184–188.

O'Connell, M., and Garner, A. (1983). Radiation-induced generation and properties of lipid hydroperoxide in liposomes. *International Journal of Radiation Biology*, 44, 615–625.

Onderdelinden, D., and Strackee, L. (1974). ESR as a tool for the identification of irradiated material. The identification of irradiated foodstuffs. pp 127–140. Luxembourg: *Commission of the European Communities*.

Ostrowski, K., Dziedzic-Goclawska, A., and Stachowicz, W. (1980). Radiation induced paramagnetic entities in tissue mineral and their use in calcified tissue research. In: Pryor, W. (Ed.), *Free Radicals in Biology* (Vol. 4). pp. 321–344. New York: Academic Press.

Ostrowski, K., Dziedzic-Goclawska, A., Stachowicz, W., and Michalik, J. (1974). Accuracy, sensitivity and specifity of electron spin resonance analysis of mineral constituents of irradiated tissues. *Annals of the New York Academy of Sciences*, 238, 186–201.

Patterson, R. L., and Stevenson, M. H. (1995). Irradiation-induced off-odor in chicken and its possible control. *British Poultry Science*, 36, 425–441.

Polonia, I., Esteves, M. P., Andrade, M. E., and Empis, J. (1995). Identification of irradiated peppers by electron spin resonance, thermoluminescence and viscosity. *Radiation Physics and Chemistry*, 46, 757–760.

Raffi, J. J., and Stocker, P. (1996). Electron paramagnetic resonance detection of irradiated foodstuffs. *Applied Magnetic Resonance*, 10, 357–373.

Raffi, J. J., Agnel, J. P. L., Buscarlet, L. A., and Martin, C. C. (1988). Electron spin resonance identification of irradiated strawberries. *Journal of Chemical Society*, 84, 3359–3362.

Raffi, J. J., Delincée, H., Marchioni, E., Hasselmann, C., Sjoberg, A. M., Leonardi, M., Kent, M., Bogl, K.W., Schreiber, G., Stevenson, H., and Meier, W. (1993). *Concerted action of the community bureau of reference on the methods of identification of irradiated foods.* Final Report-EUR/15261/EN, Brussels, Belgium.

Schreiber, G. A., Helle, N., and Bőgl, K. W. (1993). Detection of irradiated food-Methods and routine applications. *International Journal of Radiation Biology*, 63, 105–130.

Schreiber, G. A., Schulzki, G., Spiegelberg, A., Helle, N., and Bögl. K. W. (1994). Evaluation of a gas chromatographic method to identify irradiated chicken, pork, and beef by detection of volatile hydrocarbons. *Journal of AOAC International*, 77, 1202–1217.

Schulzki, G., Spiegelberg, A., Bogl, K. W., and Schreiber, G. A. (1995). Detection of radiation-induced hydrocarbons in baked sponge cake prepared with irradiated liquid egg. *Radiation Physics and Chemistry*, 46, 765–769.

Scotter, S. L., Beardwood, K., and Wood, R. (1994). Limulus amoebocyte lysate test/gram negative bacteria count method for the detection of irradiated poultry: Results of two inter-laboratory studies. *Food Science and Technology Today*, 8, 106–107.

Selvan, E., and Thomas, P. (1999). A simple method to detect gamma irradiated onions and shallots by root morphology. *Radiation Physics and Chemistry*, 55, 423–427.

Serway, R. A., and Marshall, S. A. (1967). Electron spin resonance absorption spectrum of $CO_3^-$and CO_3^- molecule-ion in irradiated single-crystal calcite. *Journal of Chemical Physics*, 47, 868–869.

Sin, D. W. M., Wong, Y. C., Yao, M. W. Y., and Marchioni, E. (2005). Identification and stability study of irradiated chicken, pork, beef, lamb, fish and mollusk shells by electron paramagnetic resonance (EPR) spectroscopy. *European Food Research and Technology*, 221, 684–691.

Soika, C., and Delincée, H. (2000). Thermoluminescence analysis for detection of irradiated food -luminescence characteristics of minerals for different types of radiation and radiation doses. *LWT-Food Science and Technology*, 33, 431–439.

Spiegelberg, A., Schulzki, G., Helle, N., Bögl, K. W., and Schreiber, G. A. (1994). Methods for routine control of irradiated food: Optimization of a method for detection of radiation-induced hydrocarbons and its application to various foods. *Radiation Physics and Chemistry*, 43, 433–444.

Stefanova, R., Vasilev, N. V., and Vassilev, N. G. (2011). ^{1}H-NMR Spectroscopy as an Alternative Tool for the Detection of γ-ray Irradiated Meat. *Food Analytical Methods*, 4, 399–403.

Stevenson, M. H., and Gray, R. (1995). The use of ESR spectroscopy for identification of irradiated food. *Annual Reports on NMR Spectroscopy*, 31, 123–142.

Stevenson, M. H., and Stewart, E. M. (1995). Identification of irradiated food - The current status. *Radiation Physics and Chemistry*, 46, 653–658.

Stevenson, M. H., Crone, A. V. J., and Hamilton, J. T. G. (1990). Irradiation detection. *Nature*, 344, 202–203.

Stewart, E. M., and Kilpatrick, D. J. (1997). An international collaborative blind trial on electronspinresonance (ESR) identification of irradiated crustacean. *Journal of the Science of Food and Agriculture*, 74, 473–484.

Stewart, E. M., McRoberts, W. C., Hamilton, J. T. G., and Graham, W. D. (2001). Isolation of lipid and 2-alkylcyclobutanones from irradiated foods by supercritical fluid extraction. *Journal of AOAC International*, 84, 976–986.

Stewart, E. M., Stevenson, M. H., and Gray, R. (1994). Use of ESR spectroscopy for the detection of irradiated crustacean. *Journal of the Science of Food and Agriculture*, 65, 191–197.

Stewart, E. M., and Stevenson, M. H. (1997). Identification of irradiated Norway Lobster (Nephrops norvegicus) using electron spin resonance (ESR) spectroscopy and estimation of applied dose using re-irradiation: results of an in-house blind trial. *Journal of the Science of Food and Agriculture*, 74, 469–472.

Stewart, F. A., Hoving, S., and Russell, N. S. (2010). Vascular damage as an underlying mechanism of cardiac and cerebral toxicity in irradiated cancer patients. *Radiation Research*, 174, 865–869.

Thakur, B. R., and Singh, R. K. (1994). Food irradiation. Chemistry and applications. *Food Reviews International*, 10, 437–473.

Trindade, R. A., Mancini-Filho, J., and Villavicencio. A. L. C. H. (2010). Natural antioxidants protecting irradiated beef burgers from lipid oxidation. *LWT - Food Science and Technology*, 43, 98–104.

Tsoulfanidis, N. (1995). *Measurement and Detection of Radiation*. second ed. New York: Taylor and Francis.

Vandenberghe, D., De Corte, F., Buylaert, J. P., Kučera, J., and Van den haute, P. (2008). On the internal radioactivity in quartz. *Radiation Measurements*, 43, 771–775.

Variyar, P. S., Chatterjee, S., Sajilata, M. G., Singhal, R. S., and Sharma, A. (2008). Natural existence of 2-alkylcyclobutanones. *Journal of Agricultural and Food Chemistry*, 56, 11817–11823.

Verbeek, F., Koppen, G., Schaeken, B., and Verschaeve, L. (2007). Automated detection of irradiated food with the comet assay. *Radiation Protection Dosimetry*, 128, 421–426.

Villavicencio, A. C. H., Filho, J. M., Hartmann, M., Ammon, J., and Delincée, H. (1997). Formation of hydrocarbons in irradiated Brazilian beans: gas chromatographic analysis to detect radiation processing. *Journal of Agricultural and Food Chemistry*, 45, 4215–4220.

von Sonntag, C. (1987). *The chemical basis of radiation biology*. London: Taylor and Francis.

Walsh, J. T., Bazinet, M. L., Kramer, R. E., and Bresnick, S. R. (1965). Hydrocarbons in irradiated beef and methyl oleate. *Journal of the American Oil Chemists' Society*, 42, 57–58.

WHO (1999). High-dose irradiation: Wholesomeness of food irradiated with doses above 10 kGy. Report of a joint FAO/IAEA/WHO study group. *WHO technical report series 890.* WHO, Geneva, Switzerland.

Winquist, F., Arwin, H., Lund, E., Forster, R., Day, C., and Lundstrom, I. Screening of irradiated tomatoes by means of an electronic nose. Proceedings of the Transducers '95, Eurosensors IX, The 8th International *Conference on Solid-state Sensors and Actuators, and Eurosensors IX*. Stockholm, Sweden; June 25–29, 1995: 691–699.

Yordanov, N. D., Aleksieva, K., Dimitrova, A., Georgieva, L., and Tzvetkova, E. (2006). Multifrequency EPR study on freeze-dried fruits before and after X-ray irradiation. *Radiation Physics and Chemistry*, 75, 1069–1074.

Yordanov, N. D., and Aleksieva, K. (2009). Preparation and applicability of fresh fruit samples for the identification of radiation treatment by EPR. *Radiation Physics and Chemistry*, 78, 213–216.

Zoumpoulakis, P., Sinanoglou, V. J., Batrinou, A., Strati, I. F., Miniadis-Meimaroglou, S., and Sflomos, K. (2012). A combined methodology to detect γ-irradiated white sesame seeds and evaluate the effects on fat content, physicochemical properties and protein allergenicity. *Food Chemistry*, 131, 713–721.

In: Ionizing Radiation
Editors: Eduard Belotserkovsky and Ziven Ostaltsov
ISBN: 978-1-62257-343-1

Chapter 2

MAMMALIAN NEVER-IN-MITOSIS-RELATED KINASE 1 IN CONTROL OF FAITHFUL CHROMATID SEGREGATION

Yumay Chen[1], Randy Wei[1], Phang-Lang Chen[1] and Daniel J. Riley[2]
[1]University of California, Irvine, California, US
[2]University of Texas Health Science Center at San Antonio, Texas, US

ABSTRACT

Protein kinases are fundamental participants in the response to DNA damage from ionizing radiation and other insults. The molecular roles of the PI3-like kinases ATM and ATR have been well characterized in the cascade of events that detect damaged DNA, activate cell cycle checkpoints, orchestrate and amplify mediators of the response, and ensure that damaged DNA is repaired before cells divide. Another mammalian protein kinase, NEK1 (never-in-mitosis related kinase 1), has similarly important but distinct roles in DNA damage responses. Studies *in vitro*, in cells, and in animals indicate that NEK1 functions uniquely as a sensor and mediator of the response to DNA damage. NEK1 is important for limiting cell death after DNA damage, activating S-phase and mitotic checkpoints properly, ensuring faithful chromosome segregation, and preventing specific neoplastic diseases. Data suggest that NEK1 deserves to be investigated further in exploring the mechanisms that lead to aneuploidy, aberrant cell death, and uncontrolled proliferation in human diseases such as kidney cancer, lymphomas, polycystic kidney disease, and bone diseases.

1. INTRODUCTION

To maintain the stability of the genome, cells must safeguard the DNA passed on to daughter cells. Any DNA damage generated by exogenous or endogenous agents must be recognized through efficient detection of the DNA lesions, and repaired with perfect fidelity prior to cell division. DNA damage generates signals that normally arrest cell cycle progression until the damaged DNA is repaired. The detection and repair process occurs at several key checkpoints before and during mitosis or meiosis. Bypassing one or more of the checkpoints, or failure of DNA damage response and repair mechanisms to correspond precisely with cell cycle checkpoints, allows damaged DNA to be passed on to the daughter cells. Cells with severe and irreparably damaged DNA suffer catastrophic mitotic errors, such that they cannot survive, while selected cells with more subtly damaged DNA might gain proliferative advantage, clonally expand, and become neoplastic.

The DNA damage-repair signaling pathway is modulated by group of precisely choreographed protein kinases. The ATM (ataxia-telangiectasia mutated) and ATR (ATM-related) kinases are the best examples of these proteins. The molecular details of their activation by DNA damage, and their cascading interactions with other kinases, have begun to explain molecular pathways important in human disease syndromes characterized by hypersensitivity to radiation, susceptibility to cancers, and premature aging. The roles of other kinases having crucial functions in guarding the genome have been discovered more recently. Screening for genes that affect cell division or that are aberrantly regulated in human cancers led to the discovery of mammalian never-in-mitosis related (NEK) and aurora kinases. These novel kinases are being shown to be important for cell cycle regulation, especially for events related to the G2/M transition and proper chromosome segregation during mitosis.

In this chapter, the unique role of NEK1 protein kinase in limiting damage after injury, including ionizing and UV radiation, will be presented. We will compare NEK1 with other never-in-mitosis related kinases, and show how it functions in a DNA signaling pathway independent of the canonical DNA damage pathway kinases ATM and ATR. The functions of NEK1 in the primary cilium-centrosome complex and in the events of mitosis will also be examined and related to the accelerated cell death, aneuploidy, and neoplastic diseases that result when NEK1 is inactivated. Finally, we will return to disease syndromes – cancers, polycystic kidney disease, and susceptibility to radiation-induced and oxidative injury – that may be targets for drugs designed to affect aberrant NEK1 functions, like drugs that specifically Aurora kinases are already being studied to treat cancer.

2. NIMA-Related Protein Kinases

In early 1970s, R. Morris and colleagues used the multicellular filamentous fungus *Aspergillus nidulans* as a model to perform the genetic analysis of cell cycle regulation. After mutagenesis, they screened hundreds of temperature sensitive mutants for their defect in cell cycle progression. They classified them as either *bim* mutants, for those blocked in mitosis, displaying condensed chromosomes and mitotic spindles, or *nim* mutants, for those that were never in mitosis due to interphase arrest. In the mid-1980s, the gene responsible for nimA mutant phenotype was cloned and designated NIMA. It was subsequently identified as a serine/threonine protein kinase. Through characterization of different nimA mutants, the essential functions of NIMA emerged. NIMA expression increases in response to DNA damage. NIMA regulates G2-M phase progression, functions as a kinase for acidic proteins, and is important for orderly mitotic events including spindle organization and proper formation of the nuclear envelope in fungi [1-3]. Early studies suggested that NIMA is required during mitosis in a “limited capacity”, as part of a cell cycle checkpoint, either one involved in making sure that DNA is properly replicated, or in assuring that damaged DNA is repaired before it's replicated [2]. Several experiments have suggested that a mitotic NIMA pathway is conserved in higher eukaryotes. First, overexpression of NIMA in *Aspergillus* results in a premature and cdc2-independent chromosome condensation, a phenomenon also observed when NIMA was expressed in yeast or human cells. Next, kinase-inactive forms of NIMA also delayed mitotic entry in human cells, presumably by interfering with the function of an endogenous protein kinase or conserved substrates.

Mammalian NIMA-related kinases (NEKs or NRKs) were identified broadly based on their sequence similarity to the catalytic domain of NIMA. At least 11 NEK protein kinases have since been identified based on their similar catalytic domain to NIMA (Figure 1). Only a few have C-terminal domains similar to that of NIMA, and these are less well characterized. NEK2 has been best characterized to date. It has been shown to have a crucial role in maintenance and regulation of chromosomes, as well as in the spindle assembly checkpoint. NEK6 and NEK7, together with NEK9 have been implicated in regulating mitotic progression [4, 5]. NEK8, like NEK1, has been linked genetically to a form of polycystic kidney disease; it localizes to the primary cilium of each cell where it functions to anchor mitotic centrosomes. NEK10 affects extracellular signal-regulated kinase 1/2 (ERK1/2) signaling to maintain G2/M arrest in response to UV irradiation. NEK11 has been linked to CDC25A degradation in response to DNA damage and is a substrate of the cell cycle checkpoint kinase CHK1. Almost all the NEKs/NRKs are involved in mitotic regulation or stress response as NIMA is. In this chapter, the focus will be the established and emerging functions of NEK1. An excellent review the molecular biology of other NEKs (NRKs) already has been published [6].

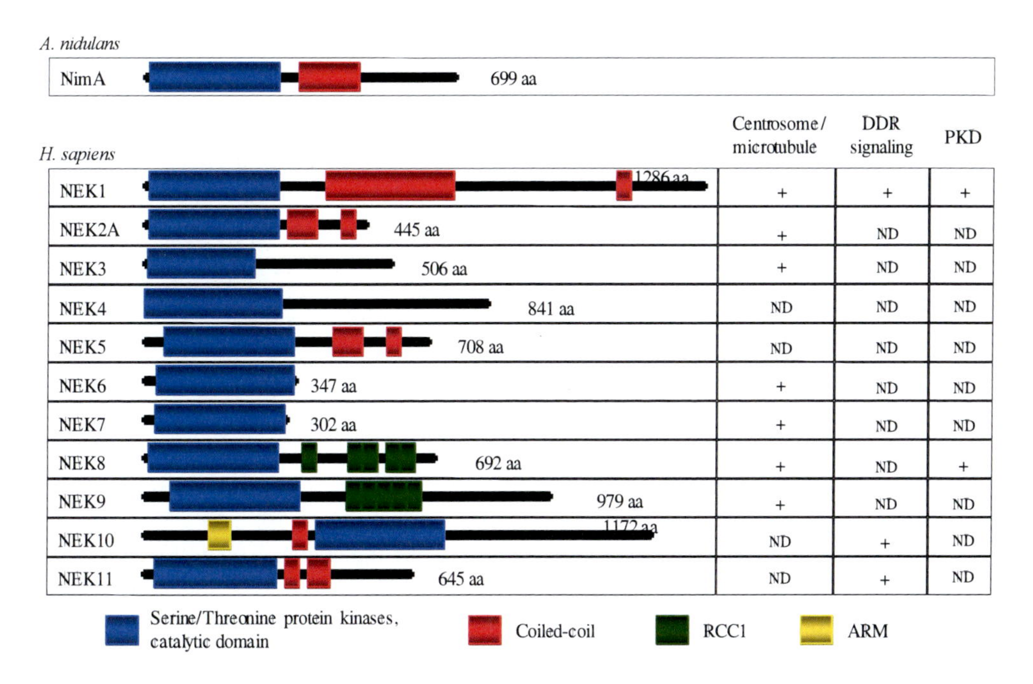

Figure 1. Alignment of the structural features of NIMA-related kinases. RCC1, RCC1-like domain; ARM, armadillo repeats; DDR, DNA damage response; PKD, polycystic kidney disease. ND, not determined.

2.1. Cloning of NIMA-Related Protein Kinase 1

NEK1 was the first cloned mammalian NIMA ortholog after being identified by screening a mouse cDNA expression library using anti-phospho-tyrosine antibodies [7]. It was originally identified as a 774 amino acid protein with 41% homology to NIMA in the N-terminal catalytic domain. This kinase domain of NEK1, when expressed from bacteria, had dual serine-threonine and tyrosine kinase activity *in vitro*. After later identifying an error in the original nucleotide sequence, the murine *Nek1* gene was actually shown to encode a protein with 1204 amino acids and a molecular mass of 160 kDa. NEK1 shares structural similarity to NIMA, a catalytic domain at N-terminal and a coiled-coil domain. At least 5 different isoforms have been identified, with the longest being 1286 a.a. [8]. It remains to be determined whether all the 5 isoforms are expressed by polymorphic alleles of the same *NEK1* and they express the functional, wild-type alleles in cells.

2.2. Dynamic Subcellular Localization of NEK1

Using specific anti-NEK1 antibodies, as well as GFP-tagged expression constructs, NEK1 was discovered to be primarily a cytosolic protein in resting, unstressed cells [9, 10]. Lesser portions of cellular NEK1 also fractionate with mitochondria and plasma membranes. After cellular injury, including DNA damage and oxidative stress, NEK1 expression is upregulated in cytoplasm and a substantial portion of it appears in nuclei at DNA damage sites. At this time, it's not known whether the upregulation is at level of transcription, RNA stability, or protein degradation. The primary NEK1 sequence indicates that the protein has two nuclear localization sequences (NLS) and one export sequence (NES) in its C-terminal domain. Deletion analysis has demonstrated the requirement of both NLS for nuclear import, and of the C-terminal NES for retention in cytoplasm. When the nuclear form of NEK1 is overexpressed in cells, the result is abnormal chromatin condensation, a phenotype similar to that seen with overexpression of NIMA. A portion of NEK1 is also found in epithelial cells at the basal body of the primary cilium [11-13], an organelle important for sensing external mechanical signals and for proper planar cell polarity as cells divide along a directional axis. The emerging role in primary cilium-centriole-centrosome complex in the genesis of aneuploidy in cancer, polycytsic kidney disease, and other diseases is discussed below (Section 4).

3. NEK1 in DNA Damage Response and Repair

DNA damage initiates a complex signaling network that rapidly senses the lesions and organizes response and repair proteins into macromolecular nuclear foci. This network activates a series of checkpoints that temporarily halt progression of the cell cycle and prevent the cell from duplicating its DNA or from undergoing mitosis until DNA can be assessed and repaired. For prompt and accurate DNA repair, signals must be conveyed rapidly and precisely. ATM, a member of the phosphatidylinositol 3-kinase-like kinase (PIKK) family, is activated at the site of a double strand DNA breaks (DSBs). Via phosphorylation of key substrates, ATM triggers a cascade of signals that sets up and amplifies multiple downstream pathways, including those that modulate DNA repair and cell cycle checkpoints [14].

The ATM and Rad3-related kinase (ATR) appears to be even more fundamental in DNA damage sensing and repair than ATM, since homozygous mutations of *ATR* have not been found in humans and since biallelic *Atr* inactivation in mice is lethal [15]. ATR has similar and intersecting downstream targets as ATM. Whereas ATM is functions primarily in response to DSBs, ATR is primarily activated by DNA replication intermediates [16, 17]. ATR is thought to be the more important upstream PI3K for signaling and repairing UV radiation- and nucleoside analog-induced DNA damage, both of which cause stalled replication forks. Until recently, ATM and/or ATR were believed to be the crucial, proximal signaling molecules in all forms of DNA damage sensing and repair.

The role of NEK1 in DNA damage sensing was first discovered NEK1-deficient cells to be much more sensitive to the effects of ionizing radiation (IR)-induced DNA damage than otherwise identical wild type cells [10]. Cells with an inactivating mutation of murine *Nek1*, as well as in cells with NEK1 expression is silenced by RNA interference, don't repair damaged DNA as efficiently as control cells. Moreover, the expression and kinase activity of NEK1 are quickly upregulated in cells treated with IR. Very early after IR, at the same time that kinase activity is upregulated, a portion of NEK1 consistently redistributes in cells from cytoplasm to nucleus, specifically in discrete foci at sites of DNA damage. At these nuclear foci, NEK1 co-localizes with key proteins involved very early in the response to IR-induced DNA double strand breaks (DSBs), including γ-H2AX and MDC1/NFBD1. The response to DNA damage is not limited to IR since NEK1 also localizes to DNA damage sites induced by alkylating agents, UV, crossing linking agents, and oxidative injury [9] (Figure 2).

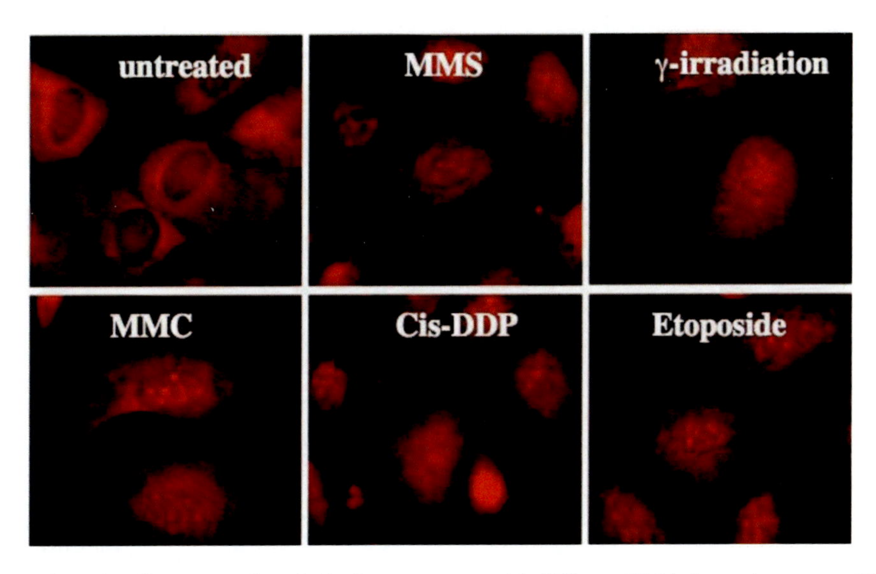

Figure 2. Nek1 relocalizes to nuclear foci after treatment with different DNA damaging agents HK2 cells in exponential growth phase on coverslips were treated with γ-radiation [ionizing radiation or IR); methylmethane sulfonate (MMS, a DNA alkylating agent), mitomycin-c, a pyrimidine cross-linking agent; etoposide, a topoisomerase II inhibitor; or cisplatin (cis-DDP), another pyrimidine cross-linker. 1 hour later, cells were fixed with 4% neutral buffered formalin and then incubated with rabbit polyclonal anti-NEK1 primary antibodies and fluorescence-tagged anti-rabbit IgG secondary antibodies.

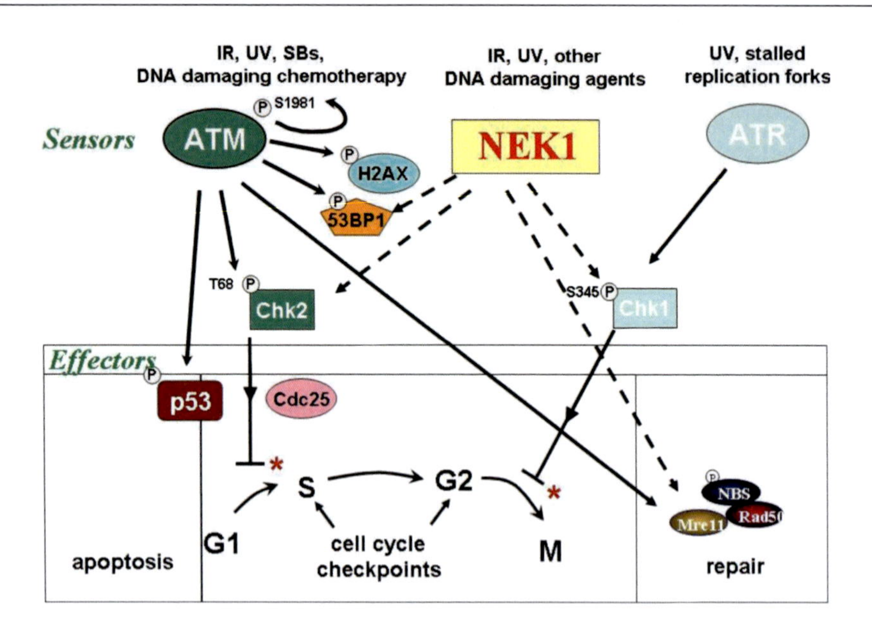

Figure 3. Schematic representation of Nek1 as a sensor kinase in the DNA damage response. Dashed arrows represent potential NEK1 substrates in known DNA damage response pathways.

NEK1 seems to oversee DNA damage responses in some ways like ATM and ATR, but it is unique and independent of both of these canonical kinases [18]. In ATM or ATR deficient cells, NEK1 expression, kinase activity, and localization to DNA damage nuclei foci are normal. Furthermore, key ATM and ATR activities are as intact in NEK1-deficient cells as they are in wild type cells. In NEK1-deficient cells, however, the crucial checkpoint kinases CHK1 and CHK2 (usually direct targets for phosphorylation by ATM or ATR) are not activated properly after defined DNA damage. Consequently, NEK1-deficient cells have defective G1/S and M-phase checkpoints. Failure of proper checkpoint responses has serious consequences: NEK1-deficient cells fail to repair damaged DNA after relatively low dose DNA damage, and they ultimately develop chromatid breaks and polyploidy in the daughter cells that survive.

How does lack of NEK1 prevent ATM and ATR from activating the downstream kinases CHK1 and CHK2? Like cells and animals with inactivated ATM, cells and animals with inactivating *NEK1* mutations are viable. NEK1 must therefore play a regulatory rather than an essential role in the cell division cycle. As a cytosolic kinase, NEK1 may be required for efficient cytoplasm to nucleus transport of CHK1 and/or CHK2. If newly synthesized CHK1 or CHK2 were not able to enter the nucleus, it could not be activated by ATM or ATR, and thus could not function properly in checkpoint control. NEK1 could activate CHK1 and CHK2 by two possible pathways. It could activate them directly if NEK1 phosphorylates a precursor site. Chk1 and Chk2 each have consensus phosphorylation sites for NEK1. Phosphorylation by NEK1 may be required for CHK1 and/or CHK2 nuclear import before ATM or ATR can phosphorylate other activating sites. Alternatively, NEK1 may regulate CHK1 and CHK2 indirectly through effects on other kinases or phosphatases that are required for CHK1 and CHK2 to interact with ATM or ATR. NEK1 could even be a redundant kinase, more important stochiometrically in response to cellular and DNA injury, such that cells without NEK1 develop problems only if stressed. In such a scenario, ATM, ATR, or other kinases would sufficient in unstressed states, but NEK1 would be required in response under conditions of oxidative stress.

Yeast two-hybrid screening has identified proteins that interact with NEK1. Several proteins involved in DNA damage response and repair pathways, such as ZBRK1, ATRX, Mre11A, and 53BP1, were found to be NEK1-interacting proteins [19]. Our unpublished results also indicate that NEK1 interacts with several other proteins involved in DNA damage response signaling, some upstream of CHK1 and CHK2, and at least one [Mre11] involved in the effector arm of DNA damage repair. The details by which NEK1 regulates these proteins remain to be examined. NEK1 may be a synergistic activator of key substrates, or an independent transducer of DNA damage signals to downstream responders (Figure 3). It may phosphorylate downstream substrates, some of the same substrates targeted by ATM and ATR, in a hit-and-run manner, perhaps differentially depending on the stimulus and the phase of the cell cycle. Further analysis of potential downstream targets of NEK1 will help to extend our understanding of the DNA damage signaling pathway and repair mechanism, and will help in the design of drugs that exploit NEK1's functional interactions for the potential treatment of neoplastic diseases.

4. The Primary Cilium-Centrosome Complex and NEK1 in Mitosis

Primary cilia, antenna-like organelles extruding from the luminal surface of all cells lining biological vestibule structures, are distinct from motile cilia. They are believed to function as mechanosensors or chemoreceptors for directional flow and for transmitting luminal signals into the cell [20-22]. Primary cilia are made up of a 9 + 0 microtubule-based structure (the axenome) and the surrounding ciliary membrane. Centrosomes function as the microtubule-organizing centers in animal cells and the center for microtubular spindle. Both primary cilium and centrosome are composed of centrioles. Centrioles are barrel-shaped structures that play a central role in the formation of centrosomes, cilia, and flagella. In cycling cells, centrosomes, composed of two centrioles, coordinate spindle pole formation during mitosis. In quiescent or interphase (G1 phase) cells, the centrosome migrates to the cell surface, whereupon the mother centriole forms a basal body that nucleates a primary cilium. The transition from primary cilium in the quiescent stage of a cell to centrosomes in the cycling stage is coordinately regulated.

4.1. Aneuploidy

NEK1 has been identified subcellularly at the base of primary cilia in epithelial cells [13, 23]. It has also been found to interact with some key components that function in the primary cilium. The motor transport protein, KIF3A, was found as a NEK1-interacting partner in the same screens that discovered interactions with DNA damage response proteins [19]. As the major component of kinesin 2 complex, KIF3A transports cargo in an anterograde direction toward the tip of the primary cilium [24]. Membrane cargo is first loaded into a vesicle at the Golgi apparatus and then transported to the basal body by dynein 1, where the cargo fuses with the cilia membrane. This primary cilium membrane-bound cargo is then transported along the ciliary length by KIF3A and dynein 2. Mutation of murine Kif3a has been found to affect the function of the cilium [25, 26]. NEK1 may regulate KIF3A activity by phosphorylation. Upon this specific phosphorylation, KIF3A might be able to recognize the cilia membrane bound cargo and transport it to the tip of the cilium. Without phosphorylation by NEK1, KIF3A may fail to transport the cargo properly, such that directional flow signals will not be transduced properly.

The primary cilium, present in interphase, is directly related the mechanics of mitosis in the same cells. A cell's primary cilium at its base is attached to the centrosomes, which will form microtubular organizing centers or spindle poles during mitosis [27]. Like almost all proteins that cause polycystic disease when inactivated, and like many proteins involved in the abnormal mitoses characteristic of cancers, NEK1 localizes at the basal body of primary cilium-centrosome complex [28]. Cells deficient in NEK1 have few normal cilia, but instead long and branched cilium-like structures. With only a few passages in culture, *NEK1 -/-* cells develop primary cilia that vary significantly in length and in number [12, 13]. These cells also develop grossly abnormal mitotic phenotypes: multiple spindle pole bodies, spindles with disparate sizes and shapes, disorganized microtubules arranged in multiple directions other than orthogonal ones, lagging chromosomes and bizarre, incomplete cytokinesis. Observations to date thus indicate that NEK1 functions to maintain orderly progression of the cell division cycle. When it is missing or inactivated, the result is aberrant mitosis.

Cells that divide without functional NEK1 develop major defects in mitotic spindle function, thus compromising the ability to segregate chromosomes faithfully to the two daughter cells, and resulting in chromosome rearrangements and aneuploidy [29, 30]. As discussed earlier, *NEK1 -/-* cells suffer from defective DNA damage response and repair. As a consequence of failure to activate checkpoints properly after DNA damage, *NEK1 -/-* cells frequently have aberrant mitotic spindles with lagging and misaligned chromosomes. The resulting cells often contain micronuclei, fragmented nuclei, or hollow nuclei. Only a small subset of these cells gains a growth advantage and continues to proliferate, whereas the majority of cells, especially those with too few chromosomes, are deleted by apoptosis. Polyploid cells gaining growth *advantages* eventually become the survivors. They proliferate without proper regulation, as malignant cells do. Polyploid *NEK1 -/-* cells are particularly remarkable in that they do not always have integer multiples of n chromosomes. Instead, they often have pieces of chromosomes and DNA content between 2n and 4n, or between 4n and 6n, as well as micronuclei, multiple nuclei of different sizes, or hollow nuclei in interphase cells. These unique phenotypes suggest that NEK1 deficiency affects not only cytokinesis, but also affects sister chromatid pairing and chromosomal rearrangements like translocations, losses, and losses with reduplications.

Mutation or inactivation of NEK1 may lead to defective mechanosensing by the primary cilium, and then to aberrant mitosis such that subsequent division of the cells occurs without proper planar polarity. Alternatively, defective DNA damage sensing and repair, bypassing of mitotic checkpoints, and disordered mitoses could be the primary problems in the setting of NEK1 deficiency. The abnormal and supernumerary primary cilia would then be the consequences of the inequitable distribution of centrosomes resulting from faulty mitoses. We suspect the latter scenario, since NEK1 clearly functions in regulating DNA damage sensing and repair after injury, and because its homologs have fundamental functions in mitosis and cell cycle progression. The aneuploid phenotype of *NEK1 -/-* cells suggests that NEK1 may have a direct role in the primary cilium-centrosome complex to regulate faithful mitosis and to ensure genomic stability. A different, more indirect role for NEK1 in resting cells (e.g., regulation of cilioproteins to ensure proper signal transduction from the primary cilium) cannot be ruled out.

5. Life and Death: NEK1's Role in Apoptosis

When cytotoxic and genotoxic stresses cause irreparable mitochondrial and DNA damage, the damaged cells are removed before passing genetic mutations on to subsequent generations of cells, often without triggering an inflammatory response in neighboring cells. This form of intrinsic cell death is unique in many ways from strictly defined apoptosis, i.e., the type that occurs during development, and may stem from a limited form of ATP depletion without frank necrosis [31]. Abrupt collapse of mitochondrial membrane potential (MMP) through opening of the mitochondrial permeability transition pore (MPTP) and subsequent mitochondrial permeabilization are seminal events in oxidative injury- and DNA damage-induced cell death. Several factors have been shown to open the MPTP, such as low intracellular pH, relative paucity of ATP, and Ca^{2+} overload. The expression of pro-apoptotic and anti-apoptotic members of the BCL2 family of proteins also been shown to modulate the status of MPTP [32, 33]. Through interaction with VDAC1 at the outer mitochondrial membrane, the balanced expression of these two groups of proteins regulates the MPTP, and in turn modulates programmed and induced cell death after injury.

Controversy exists about the role of specific components of the MPTP in all forms of intrinsic apoptosis and about the details by which the mitochondrial pores open or remain open in apoptotic versus necrotic cell death from different stimuli [34, 35]. One paradigm holds that the mitochondrial apoptosis cascade is initiated when cytochrome c exits through a pore composed of VDAC1, the inner mitochondrial membrane protein ANT (adenine nucleotide translocator), and the inner mitochondrial membrane protein cyclophilin D (cypD) [36]. Other proposed mechanisms suggest that the pore in VDAC is too narrow to allow passage of a macromolecule as large as cytochrome c, the primary function of VDAC in pro-apoptotic conditions is to regulate ATP flux by closure rather than by opening, cytochrome c release from the mitochondrial intermembrane space occurs as part of a general rupture of the outer membrane [35], or that VDACs are dispensible for mitochondrial-dependent cell death [37]. Recent studies have confirmed the importance of cypD in cell death after ischemia-reperfusion/oxidative injury (necrosis more than classically defined apoptosis) [38]. The precise roles of ANT and VDAC in the MPTP are still unsettled, but abundant evidence suggests that they are involved fundamentally in initiating mitochondrial cell death.

Kinases that may regulate VDAC activity have been reported. Hexokinases may protect cells against mitochondrial apoptosis, through direct interaction with VDAC1, to keep the channel closed and to prevent cytochrome c leakage from mitochondria [39-41]. Hexokinases are thought to allow continuous ATP flux, such that mitochondria are provided energy continuously for phosphorylation of glucose, but they do not phosphorylate VDAC directly. Candidate kinases for regulating VDAC directly have included protein kinases A and C-epsilon [42, 43], but kinase-substrate interactions for PKA or PKC-ε have never been demonstrated rigorously.

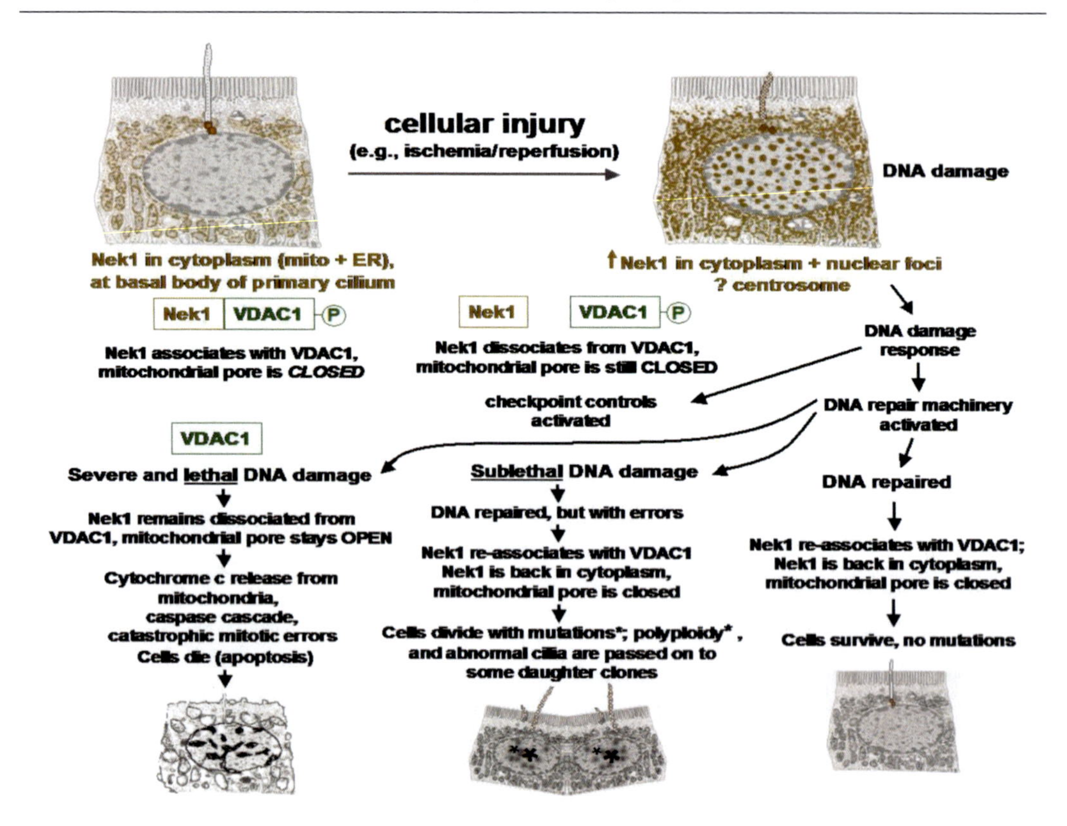

Figure 4. Role of NEK1 in limiting mitochondria-mediated cell death after injury that includes DNA damage. NEK1 participates in DNA damage responses and regulates the open-closed configuration of VDAC1 by phosphorylation.

NEK1 protein kinase, in contrast, has been shown recently by several *in vitro* and cell-based methods to interact with, phosphorylate, and regulate VDAC1. The NEK1-VDAC1 interaction was initially discovered through yeast two-hybrid screening, and confirmed by *in vitro* interaction using GST-pull down assays and *in vivo* co-immunoprecipitation [8, 44]. *In vivo*, NEK1 phosphorylates VDAC1 at serine residue 193, and thereby regulates VDAC1 channel activity and the MPTP. VDAC1 and the mitochondrial permeability transition pore open easily in the absence of functional NEK1. In the basal state of a cell, NEK1 phosphorylates VDAC1, and seems to keep the mitochondrial transition pore intact. Without functional NEK1, cells lose the VDAC1 S193 phosphorylation when injured, as by a DNA damaging agent, and die even with a DNA damage dose that would not kill identical cells that express normal amounts of functional NEK1. Furthermore, a VDAC1 mutant mimicking constitutive phosphorylation on S193 [VDAC1-Ser193→Glu], can transiently protect both *NEK1* -/- and *VDAC1* -/- cells from DNA damage-induced cell death. This data strongly support the notion that NEK1 is part of a pathway that regulates mitochondrial cell death through specific phosphorylation of VDAC1 (Figure 4).

Atomic force microscopy [AFM] was used to examine how NEK1-dependent phosphorylation regulates the opening and closing of VDAC1. Unphosphorylated, wild type VDAC1 and a VDAC1-Ser→193Ala mutant that can't be phosphorylated by NEK1 assume an open configuration. Phosphorylated, wild type VDAC1 and the VDAC1-Ser193→Glu mutant that mimics NEK1-phosphorylated VDAC1, however, are close [8, 44]. The open configuration of VDAC1 allows cytochrome c efflux, while the closed configuration of VDAC1 prevents it. These data also strongly support a direct role for VDAC1 in conducting cytochrome c and in initiating the mitochondrial-mediated cell death cascade. They also for the first time they demonstrated how a specific kinase, NEK1, regulates VDAC1 channel activity. The serine 193 residue that NEK1 phosphorylates on VDAC1 is predicted to be at a crucial site, at the junction between a C-terminal transmembrane domain and a putative cytoplasmic protein binding domain; its phosphorylation would have a significant impact on the configuration of the barrel-like channel formed by VDAC1. Two reports that used NMR and x-ray crystallography to characterize the structure of recombinant human VDAC1 in detail have identified a helical protrusion within the VDAC1 pore [45, 46]. This protrusion is comprised of N-terminal amino acids and is thought to be less stable than other regions of the VDAC1 barrel structure, such that it may switch between different conformations to control voltage gating. It is possible that the serine 193 phosphorylation of VDAC1 by NEK1 affects movement of the helical protrusion to control closing of the channel pore. It is also possible that NEK1 phosphorylation affects dimerization or oligomerization of VDAC1 [47]; such a property could also account for the large size of the pores demonstrated by AFM, which had characteristics of dimers or trimers [44].

6. Diseases Resulting from NEK1 Inactivation

For prompt and accurate DNA repair signals must be conveyed rapidly and precisely to halt cell cycle progression and to initiate DNA response-repair cascades. If these signals fail or are delayed, or if cell cycle checkpoints are bypassed, then unrepaired or misrepaired DNA can be passed onto daughter cells [48]. The damaged DNA in turn leads to defective chromatid pairing, chromosome breaks, mitotic missegregation programmed cell death if the mitotic defects are severe, and to ultimately chromosome instability in subsequent generations of the surviving daughter cells. Several syndromes associated with defective DNA damage proteins ultimately result in chromosome instability [CIN], whether the defective proteins are involved directly in centrosome or kinetochore functions during mitosis or whether they instead function indirectly in maintaining orderly cell cycle checkpoints [49]. Here, we will discuss the diseases found to date to be associated with *NEK1* mutations.

6.1. Cancer

Cancers frequently develop abnormal numbers of chromosomes and contain chromosomal rearrangements. This genomic instability generates daughter cells that die because of insufficient complements of chromosomes, as well as polyploid cells that acquire mutations favorable for uncontrolled proliferation. Genomic instability is less frequently observed in non-cancerous cells, which have competent surveillance mechanisms to monitor errors in DNA replication and chromosome segregation during mitosis, as well as the machinery to repair such damage. Dysregulation of these two important mechanisms leads to genomic instability, and ultimately to increased mutation rates and acquisition of the multiple mutations required for genesis and progression of cancer. A growing body of evidence from mouse models has linked abnormal DNA damage repair and disturbed mitotic events to the genesis and progression of cancer [50]. Study of NEK1-deficient mice in particular has advanced knowledge about how cancer and the unique neoplasia that characterizes polycystic kidney disease develop in the setting of defective DNA damage, disordered mitosis, and aberrant apoptosis.

NEK1 does not appear to be absolutely required for embryonic development in mice, which do not exhibit prominent aneuploid phenotypes during embryonic stages. Excessive apoptosis is evident, however, in the kidneys and other organs of embryonic and newborn *NEK1/kat2J -/-* mice [13]. This observation suggests that cells with abnormal chromosomes content are eliminated during development. NEK1's role in DNA damage responses may be subtle and regulatory, akin to the role associated with the key DNA damage response kinase ataxia telangiectasia mutated [ATM] [14]. *NEK1*-null mice are similar in important ways to ATM-deficient mice [18] and humans with ataxia-telangiectasia, which survive embryonic and early adult stages, but which age prematurely and develop lymphomas and other tumors later in life as they're exposed to environmental insults. ATM and Rad3-related kinase [ATR], in contrast to ATM and NEK1, is more fundamental in signaling pathways required for recognition and repair of DNA replication intermediates, and thus when inactivated in ES cells results in early embryonic lethality with fragmented chromosomes [15]. We suggest that repeated injuries or repeated doses of DNA damaging agents need to accumulate over a period of time in order to manifest gross chromosomal abnormalities and cancers late in the life of a NEK1-deficient animal. Such injuries would not occur much during embryonic development.

Since NEK1 is important for DNA damage response/repair and centrosome maintenance, the expression of sufficient amounts of NEK1 may be required for ensuring mitotic checkpoint activation, precise mitotic chromosomal segregation, and cellular cytokinesis. In fact, NEK1 deficient cells segregate their chromosomes aberrantly and acquire a polyploid, transformed phenotype. Moreover, *NEK1 +/-* cells are more sensitive to IR than cells from wild type littermates. By middles age [15-18 months], nearly 90% of *NEK1 +/-* kat2J mice develop tumors derived from B-cell, T-cell, and unclassifiable lymphocyte lineages [52]. Similar lymphoid tumors, but somewhat more restricted to T-cells, have been described in ATM-null mice [51, 53-55]. The high cumulative incidence of lymphoid tumors in *NEK1 +/-* mice suggests that low-level expression of NEK1 in cells expressing from a single allele is not sufficient to safeguard the genome and prevent chromosome instability. As is the case for most tumor suppressor gene products, loss of NEK1 expression is observed in the tumor. FACS analysis of lymphoid tumors and splenic lymphocytes from older *NEK1 +/-* mice showed cells with aberrant DNA content, not only polyploid cells with 6n and 8n DNA, but many cells with non-integer DNA content, characteristic of chromosome pieces from aberrant chromosome segregation. In this regard, the lymphoid tumors from *NEK1 +/-* mice were similar to the *NEK1 -/-* cells in culture. These animal studies revealed the importance of NEK1 in suppressing tumors. No published studies to date that have implicated NEK1 mutations in the pathogenesis of human tumors, but examining differential NEK1 expression in different human cancers should help to determine whether chromosome instability observed in cancers can be attributed to loss of NEK1 activity. More thorough examination of human tissues might also determine whether NEK1 is a prognostic marker or target for treatment of specific cancers. Further studies on whether diminished NEK1 expression leads to tumor formation in humans should be explored.

6.2. Polycystic Kidney Disease

Polycystic kidney disease [PKD] represents a unique form of neoplasia [56]. It is the most common monogenetic form of nephropathy resulting in end stage renal disease in humans. The most common form, autosomal dominant PKD, occurs because of germ line mutation in one of two genes, *PKD1* or *PKD2*, and after subsequent environmental interactions and insults. Clinical features of ADPKD usually manifest in adulthood; they include cyst formation in any portion of the nephron, hepatic cysts, colonic diverticulosis, and less frequently life-threatening cerebral arterial aneurysms and cardiac valvular abnormalities. A number of biological functions are altered in ADPKD, ranging from increased deposition of extracellular matrix, to dysregulated cellular proliferation and differentiation, to genomic instability and increased apoptosis. Tubular epithelial cells in ADPKD are thought to be arrested in their differentiation program and therefore to behave like immature precursors. Unusual apoptosis has been well documented in the kidneys of patients with ADPKD, in the cells lining cysts as well as in ostensibly unaffected renal parenchymal cells [57]. Excessive cell death is believed to occur during cystogenesis because of unusually high rates of somatic mutation [as high as 1 in 100,000 cells] in the cystic organs [58]. Polycystic kidneys and livers contain thousands of individual cysts, each lined by clones of cells with unique, second-hit *PKD1* or *PKD2* mutations in addition to the first, germ line hit [59]. The cysts grow at varying rates, new ones develop throughout life, and the oldest cysts tend to have the most aggressive growth [60].

Several animal models of polycystic kidney disease have been generated, some unexpectedly, by deregulation of cellular proliferation or apoptosis. Such animal models include strains that up-regulate proliferation-enhancing, pro-apoptotic proteins like c-myc [61] and constitutively activated β-catenin [62], or inactivate anti-apoptotic proteins such as Bcl-2 [63-65], transcription factor AP2β [66], and the *PKD1* gene product itself, polycystin-1 [67]. Dysregulated proliferation and apoptosis, therefore, play important roles in polycystic kidney disease, just like they do on cancer.

Mutations that inactivate NEK1 generated excellent mouse models of PKD and the mechanisms that underlie its development and progression. Two independent strains [the so-called kidneys-anemia-testes mice] were developed by selectively crossing spontaneous B6 and C57BL mutants at The Jackson Laboratory. The gene underlying the observed PKD phenotypes in each strain was subsequently mapped to NEK1. The mutations found in these two strains, kat (deletion) and kat2J (insertion leading to early termination), result in the expression of inactive NEK1 kinase lacking the carboxyl terminal coiled-coil region. The kat2J strain with the more severely truncated NEK1 protein develops the more aggressive PKD. This strain was originally thought to be a model of a recessive form of PKD, with cysts limited to renal glomeruli. Detailed analysis of kat2J mice by our group, however, clarified the PKD phenotype. Using different markers specific for tubular subtypes, we found that cysts in the NEK1-deficient mice form from all parts of the nephron [13], including Bowman's capsule surrounding glomeruli, proximal tubules, loops of Henle, and collecting ducts, just like cysts in humans with *PKD1* mutations do. We also discovered that haplo-insufficiency for murine NEK1 results in a less severe polycystic phenotype in most mice in their old age, and that cyst-lining epithelial cells lose expression of NEK1. These results suggest that loss of NEK1 expression may be important, either etiologically in cyst generation or as a consequence of early cyst development, and that stochastic inactivation of NEK1 may be required for cystogenesis in the NEK1-deficient kat2J model.

Current paradigms contend that diseases like polycystic kidney disease because of problems in the function of the primary cilium-centrosome complex [28]. The directional propagation of cells lining tubes [e.g., renal tubules and bile ducts in the liver] requires precise planar polarity as the tubes grow during development or as they repopulate the lining of established tubes. If cells growing along a tube do not divide in the proper orientation, then they form outpockets, the nascent foci of cysts. A key function of primary cilia is believed to be sensing the axis of cell division, and making sure that daughter cells move in the proper direction to line tubules and ducts.

Another key requirement for the development of PKD is multiple mitoses. In animal models in which key primary cilia proteins can be conditionally inactivated at different times, the PKD phenotype is observed only when they're inactivated while kidney cells are actively dividing [68, 69]. Such times include embryonic or early postnatal development and after the tubules are injured in such a manner that epithelial cells need to reline the tubular lumens. The key element required during all of these times is mitosis.

Oxidative or ischemia-reperfusion injury and DNA damage after full kidney development has even been described as the "third hit" required for disese progression in ADPKD [70]. Abnormalities in the survival of dividing cells [aneuploidy and apoptosis], excessive and dysregulated proliferation of a selected clones, and aberrant planar polarity all result from errors in mitosis [71, 72], precisely the same kind of errors that occur when NEK1 is inactivated. NEK1 thus seems to have similar roles in protecting against mitotic errors involved in the generation of cancers and in PKD.

These roles and the known effects of ionizing radiation in causing damage to DNA should also give doctors pause about getting too many CAT scans or any DNA-damaging chemotherapeutic agents in patients with ADPKD, lest these treatments hasten the progression of kidney disease or lead to cancers in cells that can't repair damaged DNA properly.

At least one other NIMA-related kinase has been associated with PKD. NEK8, which contains unique RCC1 repeat domains important for chromosomal condensation in other proteins, was identified as the gene locus mutated in jck (juvenile polycystic kidney) mice. These animals develop cysts primarily in their collecting tubules [73]. NEK8 localizes to the apical cytoplasm of inner medullary collecting tubules, but is mislocated in the collecting tubules of jck mice.

Expression of a dominant negative form of NEK8, based on a similar kinase-inactivating mutation in NIMA, results in a disordered actin cytoskeleton and in multiple nuclei; these findings suggest that NEK8 may serve an essential function in regulating cytoskeletal structure and chromosome segregation in a specific subset of renal cells. NEK1 and NEK8 both are involved in the pathogenesis PKD, then, and both seem to be involved in DNA damage responses. It's interesting to note that NEK1 localizes in a different subset of kidney cells, glomerular and proximal tubular epithelial cells, which derive from metanephric mesenchyme rather from ureteric bud (the embryonic origin of collecting tubules). Therefore NEK1 and NEK8 could serve similar functions in different subsets of kidney cells derived from different embryonic precursors.

6.3. Bone Disease

NEK1 mutations were found in two families with autosomal-recessive short-rib polydactyly syndrome [SRPS] Majewski type [74]. There are four types of this autosomal recessive, lethal skeletal dysplasias: SRPS I [Saldino-Noonan type], SRPS II [Majewski type], SRPS III [Verma-Naumoff type], and SRPS IV [Beemer-Langer type]. Patients with SRPS are characterized by markedly short ribs, short limbs, polydactyly, and multiple anomalies of major organs, including heart, intestines, genitalia, kidney, liver, and pancreas. Patients with SRPS II have characteristically short, oval tibias. Other genes responsible for SRPS are *IFM80* and *DYNC2H*1, both of which express proteins in the primary cilium-centrosome complex [75, 76]. Abnormal bone and cartilage phenotypes are also observed in mice with the in the kat2J *NEK1* mutation. The most easily observed phenotypes are facial dysmorphism and stunted growth of long bones [77]. Preliminary studies in our lab have shown disorganized endochondral ossification with excessive apoptosis in cells of the epiphyseal proliferative zones, which rapidly proliferate during bone elongation. These findings in NEK1-deficient mice are very similar to those observed in humans with SRPS II. Patients with SRPS II are also frequently found to have cystic kidney disease. Other reports have shown that mutations at the *PKD1* locus, in addition to being associated with cystic kidneys and liver, are sometimes associated with skeletal limb defects, including polydactyly [78]. Like PKD, bone diseases seem to stem from aberrant functions of the primary cilium-centrosome complex, with mitotic segregation errors, altered planar polarity, and aberrant apoptosis, as bones must grow or remodel in a coordinated direction and with precise timing.

CONCLUSION

Nek1 is a serine/threonine/tyrosin kinase with multiple functions that regulate a cell's response to injury, especially injury that includes DNA damage. NEK1 expression is upregulated in response to injury, and portions of it translocate from cytoplasm to nuclear foci very soon after DNA damage. Another portion localizes to the primary cilium-centrosome complex in a cell, a structure that is crucial for cells to sense directional flow, establish planar polarity when they divide along a defined axis, ultimately as the microtubular organizing centers during mitosis. Without functional NEK1, many cells have bypass or mis-time cell cycle checkpoints, fail to repair DNA properly, and suffer from errors in mitosis, including inequitable sister chromatid segregation and cytokinesis. Many of these damaged cells die, while others become polyploid and ultimately transformed. In addition to functioning in DNA damage signaling pathways and mitotic segregation events, NEK1 is also involved fundamentally in mitochondria-mediated cell death, through its regulation of voltage dependent anion channels and an injured cell's mitochondrial permeability transition.

Cells without functional NEK1 are hypersusceptible to DNA damage from ionizing radiation. Subsets of these cells that survive faulty DNA damage repair ultimately become polyploid, and can transform into neoplastic cells. The kat and kat2J strains of NEK1-deficient mice are useful for modeling several neoplastic states that mimic human diseases, including polycystic kidney disease, lymphoid cancers, and developmental bone diseases. Each of these diseases stems from regulatory problems in mitosis and programmed or injury-induced cell death. Further understanding of how NEK1 fits into the DNA damage response, and how it guards against aneuploidy, will be important for preventing the progression of somewhat predictable neoplastic diseases like PKD.

NEK1 is involved as a sensor or mediator of DNA damage responses, in ways similar to but distinct from and independent of the functions of the classical DNA damage response kinase ATM. Multifunctional NEK1 has not yet received much attention as a potential marker for injury (e.g., acute kidney injury, DNA damage) or for certain types of cancer, but it should in the future. Furthermore, drugs or small molecules that increase NEK1 expression, or that mimic interactions with some of its known substrates, might prove to be useful for limiting the adverse consequences of acute injury associated with radiation doses or genotoxic chemotherapy.

ACKNOWLEDGMENTS

This chapter was supported by grant from NIH (R01-DK067339) to Y.C.

REFERENCES

[1] Osmani A. H., McGuire S. L., Osmani S. A. Parallel activation of the NIMA and p34cdc2 cell cycle-regulated protein kinases is required to initiate mitosis in A. nidulans. *Cell* 1991; 67: 283-91.

[2] Osmani A. H., O'Donnell K., Pu R. T., Osmani S. A. Activation of the nimA protein kinase plays a unique role during mitosis that cannot be bypassed by absence of the bimE checkpoint. *Embo J* 1991; 10: 2669-79.

[3] Osmani S. A., Engle D. B., Doonan J. H., Morris N. R. Spindle formation and chromatin condensation in cells blocked at interphase by mutation of a negative cell cycle control gene. *Cell* 1988; 52: 241-51.

[4] Roig J., Mikhailov A., Belham C., Avruch J. Nercc1, a mammalian NIMA-family kinase, binds the Ran GTPase and regulates mitotic progression. *Genes Dev* 2002; 16: 1640-58.

[5] Yissachar N., Salem H., Tennenbaum T., Motro B. Nek7 kinase is enriched at the centrosome, and is required for proper spindle assembly and mitotic progression. *FEBS Lett* 2006; 580: 6489-95.

[6] O'Connell M. J., Krien M. J., Hunter T. Never say never. The NIMA-related protein kinases in mitotic control. *Trends Cell Biol* 2003; 13: 221-8.

[7] Letwin K., Mizzen L., Motro B., Ben-David Y., Bernstein A., Pawson T. A mammalian dual specificity protein kinase, Nek1, is related to the NIMA cell cycle regulator and highly expressed in meiotic germ cells. *Embo J* 1992; 11: 3521-31.

[8] Chen Y., Craigen W. J., Riley D. J. Nek1 regulates cell death and mitochondrial membrane permeability through phosphorylation of VDAC1. *Cell Cycle* 2009; 8: 257-67.

[9] Chen Y., Chen P. L., Chen C. F., Jiang X., Riley D. J. Never-in-mitosis related kinase 1 functions in DNA damage response and checkpoint control. *Cell Cycle* 2008; 7: 3194-201.

[10] Polci R., Peng A., Chen P. L., Riley D. J., Chen Y. NIMA-related protein kinase 1 is involved early in the ionizing radiation-induced DNA damage response. *Cancer Res* 2004; 64: 8800-3.

[11] Mahjoub M. R., Montpetit B., Zhao L., Finst R. J., Goh B., Kim A. C. et al. The FA2 gene of Chlamydomonas encodes a NIMA family kinase with roles in cell cycle progression and microtubule severing during deflagellation. *J Cell Sci* 2002; 115: 1759-68.

[12] Shalom O., Shalva N., Altschuler Y., Motro B. The mammalian Nek1 kinase is involved in primary cilium formation. *FEBS Lett* 2008; 582: 1465-70.

[13] Chen Y., Chiang H. C., Litchfield P., Pena M., Riley D. J. *Kidneys develop aberrantly and cysts derive from multiple different nephron segments in the Nek1-deficient kat2J mouse model of polycystic kidney disease.* Manuscript submitted 2012.

[14] Shiloh Y. The ATM-mediated DNA-damage response: taking shape. *Trends Biochem Sci* 2006; 31: 402-10.

[15] Brown E. J., Baltimore D. ATR disruption leads to chromosomal fragmentation and early embryonic lethality. *Genes Dev* 2000; 14: 397-402.

[16] Cortez D., Guntuku S., Qin J., Elledge S. J. ATR and ATRIP: partners in checkpoint signaling. *Science* 2001; 294: 1713-6.

[17] Hurley P. J., Bunz F. ATM and ATR: components of an integrated circuit. *Cell Cycle* 2007; 6: 414-7.

[18] Chen Y., Chen C. F., Riley D. J., Chen PL. Nek1 kinase functions in DNA damage response and checkpoint control through a pathway independent of ATM and ATR. *Cell Cycle* 2011;10: 655-63.

[19] Surpili M. J., Delben T. M., Kobarg J. Identification of proteins that interact with the central coiled-coil region of the human protein kinase NEK1. *Biochemistry* 2003; 42: 15369-76.

[20] Pazour G. J., Witman G. B. The vertebrate primary cilium is a sensory organelle. *Curr Opin Cell Biol* 2003; 15: 105-10.

[21] Singla V., Reiter J. F. The primary cilium as the cell's antenna: signaling at a sensory organelle. *Science* 2006; 313: 629-33.

[22] Wheatley D. N. Primary cilia in normal and pathological tissues. *Pathobiology* 1995; 63: 222-38.

[23] Mahjoub M. R., Trapp M. L., Quarmby, L. M. NIMA-related kinases defective in murine models of polycystic kidney diseases localize to primary cilia and centrosomes. *J Am Soc Nephrol* 2005; 16: 3485-9.

[24] Rosenbaum J. L., Witman G. B. Intraflagellar transport. *Nat Rev Mol Cell Biol* 2002; 3: 813-25.

[25] Lin F., Hiesberger T., Cordes K., Sinclair A. M., Goldstein L. S., Somlo S. et al. Kidney-specific inactivation of the KIF3A subunit of kinesin-II inhibits renal ciliogenesis and produces polycystic kidney disease. *Proc Natl Acad Sci U S A* 2003; 100: 5286-91.

[26] Takeda S., Yonekawa Y., Tanaka Y., Okada Y., Nonaka S., Hirokawa N. Left-right asymmetry and kinesin superfamily protein KIF3A: new insights in determination of laterality and mesoderm induction by kif3A-/- mice analysis. *J Cell Biol* 1999; 145: 825-36.

[27] Pan J., Snell W. The primary cilium: keeper of the key to cell division. *Cell* 2007; 129: 1255-7.

[28] Guay-Woodford L. M. Renal cystic diseases: diverse phenotypes converge on the cilium/centrosome complex. *Pediatr Nephrol* 2006; 21: 1369-76.

[29] Pihan G. A., Doxsey S. J. The mitotic machinery as a source of genetic instability in cancer. *Semin Cancer Biol* 1999; 9: 289-302.

[30] Sorger P. K., Dobles M., Tournebize R., Hyman A. A. Coupling cell division and cell death to microtubule dynamics. *Curr Opin Cell Biol* 1997; 9: 807-14.

[31] Kroemer G., Galluzzi L., Brenner C. Mitochondrial membrane permeabilization in cell death. *Physiol Rev* 2007; 87: 99-163.

[32] Vander Heiden M. G., Thompson C. B. Bcl-2 proteins: regulators of apoptosis or of mitochondrial homeostasis? *Nat Cell Biol* 1999; 1: E209-16.

[33] Wang X. The expanding role of mitochondria in apoptosis. *Genes Dev* 2001; 15: 2922-33.

[34] Green D. R., Kroemer G. The pathophysiology of mitochondrial cell death. *Science* 2004; 305: 626-9.

[35] Rostovtseva T. K., Tan W., Colombini M. On the role of VDAC in apoptosis: fact and fiction. *J Bioenerg Biomembr* 2005; 37: 129-42.

[36] Tsujimoto Y., Shimizu S. The voltage-dependent anion channel: an essential player in apoptosis. *Biochimie* 2002; 84: 187-93.

[37] Baines C. P., Kaiser R. A., Sheiko T., Craigen W. J., Molkentin J. D. Voltage-dependent anion channels are dispensable for mitochondrial-dependent cell death. *Nat Cell Biol* 2007; 9: 550-5.

[38] Baines C. P., Kaiser R. A., Purcell N. H., Blair N. S., Osinska H., Hambleton M. A. et al. Loss of cyclophilin D reveals a critical role for mitochondrial permeability transition in cell death. *Nature* 2005; 434: 658-62.

[39] Abu-Hamad S., Zaid H., Israelson A., Nahon E., Shoshan-Barmatz V. Hexokinase-I protection against apoptotic cell death is mediated via interaction with the voltage-dependent anion channel-1: mapping the site of binding. *J Biol Chem* 2008; 283: 13482-90.

[40] Azoulay-Zohar H., Israelson A., Abu-Hamad S., Shoshan-Barmatz V. In self-defence: hexokinase promotes voltage-dependent anion channel closure and prevents mitochondria-mediated apoptotic cell death. *Biochem J* 2004; 377:347-55.

[41] Sun L., Shukair S., Naik T. J., Moazed F., Ardehali H. Glucose phosphorylation and mitochondrial binding are required for the protective effects of hexokinases I and II. *Mol Cell Biol* 2008; 28: 1007-17.

[42] Bera A. K., Ghosh S., Das S. Mitochondrial VDAC can be phosphorylated by cyclic AMP-dependent protein kinase. *Biochem Biophys Res Commun* 1995; 209: 213-7.

[43] Saigusa A., Kokubun S. Protein kinase C may regulate resting anion conductance in vascular smooth muscle cells. *Biochem Biophys Res Commun* 1988; 155: 882-9.

[44] Chen Y., Gaczynska M., Osmulski P., Polci R., Riley D. J. Phosphorylation by Nek1 regulates opening and closing of voltage dependent anion channel 1. *Biochem Biophys Res Commun* 2010; 394: 798-803.

[45] Bayrhuber M., Meins T., Habeck M., Becker S., Giller K., Villinger S. et al. Structure of the human voltage-dependent anion channel. *Proc Natl Acad Sci U S A* 2008; 105: 15370-5.

[46] Hiller S., Garces R. G., Malia T. J., Orekhov V. Y., Colombini M., Wagner G. Solution structure of the integral human membrane protein VDAC-1 in detergent micelles. *Science* 2008; 321: 1206-10.

[47] Goncalves R. P., Buzhynskyy N., Prima V., Sturgis J. N., Scheuring S. Supramolecular assembly of VDAC in native mitochondrial outer membranes. *J Mol Biol* 2007; 369: 413-8.

[48] Kaye J. A., Melo J. A., Cheung S. K., Vaze M. B., Haber J. E., Toczyski D. P. DNA breaks promote genomic instability by impeding proper chromosome segregation. *Curr Biol* 2004; 14: 2096-106.

[49] Rajagopalan H., Lengauer C. Aneuploidy and cancer. *Nature* 2004; 432: 338-41.

[50] Schvartzman J. M., Sotillo R., Benezra R. Mitotic chromosomal instability and cancer: mouse modelling of the human disease. *Nat Rev Cancer* 2010; 10: 102-15.

[51] Barlow C., Hirotsune S., Paylor R., Liyanage M., Eckhaus M., Collins F. et al. Atm-deficient mice: a paradigm of ataxia telangiectasia. *Cell* 1996; 86: 159-71.

[52] Chen Y., Chen C. F., Chiang H. C., Pena M., Polci R., Wei R. L. et al. Mutation of NIMA-related kinase 1 (NEK1) leads to chromosome instability. *Mol Cancer* 2011; 10: 5.

[53] Elson A., Wang Y., Daugherty C. J., Morton C. C., Zhou F., Campos-Torres J. et al. Pleiotropic defects in ataxia-telangiectasia protein-deficient mice. *Proc Natl Acad Sci U S A* 1996; 93: 13084-9.

[54] Spring K., Cross S., Li C., Watters D., Ben-Senior L., Waring P. et al. Atm knock-in mice harboring an in-frame deletion corresponding to the human ATM 7636del9 common mutation exhibit a variant phenotype. *Cancer Res* 2001; 61: 4561-8.

[55] Xu Y., Ashley T., Brainerd E. E., Bronson R. T., Meyn M. S., Baltimore D. Targeted disruption of ATM leads to growth retardation, chromosomal fragmentation during meiosis, immune defects, and thymic lymphoma. *Genes Dev* 1996; 10: 2411-22.

[56] Grantham J. J. Polycystic kidney disease: neoplasia in disguise. *Am J Kidney Dis* 1990;15(2):110-6.

[57] Woo D. Apoptosis and loss of renal tissue in polycystic kidney diseases. *N Engl J Med* 1995;333(1):18-25.

[58] Peters D. J., Breuning M. H. Autosomal dominant polycystic kidney disease: modification of disease progression. *Lancet* 2001; 358 (9291):1439-44.

[59] Qian F., Watnick T. J., Onuchic L. F., Germino G. G. The molecular basis of focal cyst formation in human autosomal dominant polycystic kidney disease type I. *Cell* 1996;87(6):979-87.

[60] Chapman A. B., Guay-Woodford L. M., Grantham J. J., Torres V. E., Bae K. T., Baumgarten D. A. et al. Renal structure in early autosomal-dominant polycystic kidney disease (ADPKD): The Consortium for Radiologic Imaging Studies of Polycystic Kidney Disease (CRISP) cohort. *Kidney Int* 2003; 64(3): 1035-45.

[61] Trudel M., D'Agati V., Costantini F. C-myc as an inducer of polycystic kidney disease in transgenic mice. *Kidney Int* 1991; 39: 665-71.

[62] Saadi-Kheddouci S., Berrebi D., Romagnolo B., Cluzeaud F., Peuchmaur M., Kahn A. et al. Early development of polycystic kidney disease in transgenic mice expressing an activated mutant of the beta-catenin gene. *Oncogene* 2001; 20:5972-81.

[63] Sorenson C. M., Padanilam B. J., Hammerman M. R. Abnormal postpartum renal development and cystogenesis in the bcl-2 (-/-) mouse. *Am J Physiol* 1996; 271: F184-93.

[64] Sorenson C. M., Rogers S. A., Korsmeyer S. J., Hammerman M. R. Fulminant metanephric apoptosis and abnormal kidney development in bcl-2-deficient mice. *Am J Physiol* 1995; 268: F73-81.

[65] Veis D. J., Sorenson C. M., Shutter J. R., Korsmeyer S. J. Bcl-2-deficient mice demonstrate fulminant lymphoid apoptosis, polycystic kidneys, and hypopigmented hair. *Cell* 1993; 75: 229-40.

[66] Moser M., Pscherer A., Roth C., Becker J., Mucher G., Zerres K. et al. Enhanced apoptotic cell death of renal epithelial cells in mice lacking transcription factor AP-2beta. *Genes Dev* 1997; 11: 1938-48.

[67] Boletta A., Qian F., Onuchic L. F., Bhunia A. K., Phakdeekitcharoen B., Hanaoka K. et al. Polycystin-1, the gene product of PKD1, induces resistance to apoptosis and spontaneous tubulogenesis in MDCK cells. *Mol Cell* 2000; 6: 1267-73.

[68] Luyten A., Su X., Gondela S., Chen Y., Rompani S., Takakura A. et al. Aberrant regulation of planar cell polarity in polycystic kidney disease. *J Am Soc Nephrol* 2010;21(9):1521-32.

[69] Patel V., Li L., Cobo-Stark P., Shao X., Somlo S., Lin F. et al. Acute kidney injury and aberrant planar cell polarity induce cyst formation in mice lacking renal cilia. *Hum Mol Genet* 2008; 17: 1578-90.

[70] Zhou J. Polycystins and primary cilia: primers for cell cycle progression. *Annu Rev Physiol* 2009; 71:83-113.

[71] Battini L., Macip S., Fedorova E., Dikman S., Somlo S., Montagna C. et al. Loss of polycystin-1 causes centrosome amplification and genomic instability. *Hum Mol Genet* 2008; 17: 2819-33.

[72] Burtey S., Riera M., Ribe E., Pennenkamp P., Rance R., Luciani J. et al. Centrosome overduplication and mitotic instability in PKD2 transgenic lines. *Cell Biol Int* 2008; 32: 1193-8.

[73] Liu S., Lu W., Obara T., Kuida S., Lehoczky J., Dewar K. et al. A defect in a novel Nek-family kinase causes cystic kidney disease in the mouse and in zebrafish. *Development* 2002; 129: 5839-46.

[74] Thiel C., Kessler K., Giessl A., Dimmler A., Shalev S. A., von der Haar S. et al. NEK1 mutations cause short-rib polydactyly syndrome type majewski. *Am J Hum Genet* 2011; 88: 106-14.

[75] Dagoneau N., Goulet M., Genevieve D., Sznajer Y., Martinovic J., Smithson S. et al. DYNC2H1 mutations cause asphyxiating thoracic dystrophy and short rib-polydactyly syndrome, type III. *Am J Hum Genet* 2009; 84: 706-11.

[76] Merrill A. E., Merriman B., Farrington-Rock C., Camacho N., Sebald E. T., Funari V. A. et al. Ciliary abnormalities due to defects in the retrograde transport protein DYNC2H1 in short-rib polydactyly syndrome. *Am J Hum Genet* 2009; 84: 542-9.

[77] Upadhya P., Birkenmeier E. H., Birkenmeier C. S., Barker J. E. Mutations in a NIMA-related kinase gene, Nek1, cause pleiotropic effects including a progressive polycystic kidney disease in mice. *Proc Natl Acad Sci U S A* 2000;97(1): 217-21.

[78] Turco A. E., Padovani E. M., Chiaffoni G. P., Peissel B., Rossetti S., Marcolongo A. et al. Molecular genetic diagnosis of autosomal dominant polycystic kidney disease in a newborn with bilateral cystic kidneys detected prenatally and multiple skeletal malformations. *J Med Genet* 1993; 30: 419-22.

In: Ionizing Radiation
Editors: Eduard Belotserkovsky and Ziven Ostaltsov
ISBN: 978-1-62257-343-1

Chapter 3

THE DIFFUSION AND AGGREGATION OF INTRINSIC RADIATIVE DEFECTS IN LITHIUM FLUORIDE CRYSTALS AND NANOCRYSTALS

A. P. Voitovich
Institute of Physics, National Academy of Sciences, Minsk, Belarus

ABSTRACT

Bulk crystals and nanocrystals of lithium fluoride are irradiated by various doses of gamma rays at a temperature of 77 K. The time evolution of photoluminescence from F_2^+, F_2, F_3^+ and F_3 color centers are measured at various annealing temperatures. It has been revealed that in many cases the kinetics of the reactions, determined by mobile defects diffusion in crystals, is described by the exponential dependence. At diffusion of random walk type, on the base of such dependence distribution of the diffusion pathways travelled by the mobile defects prior to entry into reactions, has been obtained. Lifetimes of anionic vacancies υ_a and F_2^+ centers, also activation energies and coefficients of their diffusion are determined. It is found that in the bulk crystals lifetime decreases for vacancies while increases for F_2^+ centers by increasing the irradiation dose. It is also shown that, after irradiation during crystals annealing, vacancies are formed as a result of the reaction $F_2^+ + H \rightarrow \upsilon_a + Fl^-$, where Fl^- is a fluorine ion in a lattice site and H is a fluorine interstitial atom. The presence of F_1^- centers in the irradiated samples is established, and the processes, which lead to the formation of F_2, F_3^+ and F_3 centers after irradiation, are discovered. It has been found that diffusion activation energy for vacancies in nanocrystals is twice as high as compared with bulk crystals. Migration of F_2^+ centers in nanocrystals has not been detected. Presence of new type of radiative defects in nanocrystals as compared with those in bulk crystals has been experimentally revealed.

I. INTRODUCTION

The radiation-modified crystals are widely used in electronics, dosimetry and laser technology [1-3]. The problems of structures design with high spatial resolution based on radiation-induced defects have been studied [4-6]. The primary intrinsic mobile defects, which are created as a result of radiation exposure, diffuse through a crystal and enter into reactions with the localized defects. When replacement or aggregation reactions proceed restoration of lattice regularity or formation of aggregated defects take place. Thus, diffusion of mobile defects in many aspects defines the properties of the irradiated crystals. In virtue of this fact it is intensively investigated. Also, the reactions kinetics in which mobile defects participate is being studied and activation energies of their diffusion are being evaluated [7-10]. The mobile defects travel some diffusion pathways prior to entry into reaction. One of the major characteristics which define kinetics laws is distribution of diffusion pathways which mobile defects pass prior to entry into reaction according to their values. Such distribution is equal to distribution of the defects participating in a reaction according to distances between them. The researches devoted to determination of such distribution are unknown to us.

In the present chapter distribution of diffusion pathways travelled by mobile defects in a crystal prior to entry into a reaction will be presented. Distribution is being deduced on the basis of experimentally found kinetics laws of reactions in which mobile and localized defects participate during the post-radiation period. Use of the defined laws and distribution for calculation of mobile defects diffusion characteristics is illustrated. The investigations have been carried out with lithium fluoride crystals in which there is no gradient of defects concentrations. As it is accepted in literature the terminology «diffusion of vacancies and color centers» is used, though actually ions diffuse.

Among alkali halides crystals, lithium fluoride (LiF) occupies a special place because of its physical and optical properties. Color centers (radiation-induced defects) in this material were extensively investigated and found application in the realization of broad-band emitting lasers in the optical domain operating at room temperature (RT) [3]. In the optoelectronic application sub-micron periodic gratings were written inside LiF crystals by a mode-locked Ti:sapphire laser. Pure [11] and doped [12] LiF crystals are well-known dosimeter materials. Recently new radiation detectors based on microcrystalline LiF in polymer matrix for gamma and electrons high-dose dosimetry have been proposed [13], as well polycrystalline LiF films as nuclear sensor for neutrons [14]. Polycrystalline lithium fluoride (LiF ceramics) has been created [15] and the possibilities for its use have been intensively studied. Applications of lithium fluoride with color centers (CCs) require targeted variation and optimization of its properties and characteristics. In order to solve these problems, we need to know as much detail as possible about the color center formation processes.

Processes involved in formation of radiation-induced CCs in lithium fluoride have been studied for many years [7, 16–19]. The usual techniques of additive coloration are not effective in the formation of defects in these crystals. They can be colored by several types of ionizing radiations, as X-rays, γ-rays, electrons, protons, neutrons, α-particles and heavier charged ions. Each radiation source gives rise to different kinds of damaging processes. We shall use here only the γ-rays coloration. Gamma rays have a high penetration power and they lose their energy through the material along thickness of some millimeters, uniformly. It has been found that the kinetics and results of coloring LiF depend on a number of factors: the irradiation dose and irradiation conditions, the chemical purity, the mechanical and thermal history of the crystal, etc.

In this chapter we consider and compare the kinetics and processes for the formation of F_2^+, F_2, F_3^+, F_3 color centers in LiF bulk crystals and nanocrystals (hundreds of nanometers) during annealing in the post-radiation stage, after irradiation by gamma photons at liquid nitrogen temperature, i.e. at $T_{irrad} = 77$ K . (It should be noted that in designations of the centers F_n^-, F_n^+, F_n the lower index "n" is a number of positively charged anion vacancies υ_a which enter into the center; the upper indexes "+" or "–" (absence of index) mean an excess or deficiency (equality) of electrons in the center as compared with the number of vacancies.) The post-radiation period was selected for study in connection with the fact that, in this period, fewer processes participate in generation of color centers than during irradiation, and these processes are more easily isolated.

II. Samples and Study Technique

It was shown earlier [20] that the conditions imposed on irradiation of crystals intended for studying color center aggregation kinetics in the post-radiation period are quite well satisfied if the samples are irradiated by gamma photons at a temperature T_{irrad} that is lower than the temperature T_υ of vacancy mobility. Accordingly, all the samples were irradiated by gamma photons from a ^{60}Co source at liquid nitrogen temperature, $T_{irrad} = 77$ K. The irradiation temperature was below the V_k center and vacancy mobility temperatures, so that these defects were localized during irradiation. Therefore after irradiation was finished, the following defects of interest to us existed in the crystal: interstitial fluorine ions (I) and fluorine atoms (H), vacancies (υ_a), V_k centers, single-vacancy F_1 and F_1^- centers. It is hypothesized that the irradiation doses are fairly low and the probability of generating aggregate color centers during irradiation as a result of binary or higher order processes is negligibly small.

Crystalline samples for the studies were punched out from a nominally pure LiF single crystal along the {100} plane in the form of (0.7–1.5) x (5–7) x (7–10) mm^3 plates. Measurements of the transmission spectra showed that the absorption band with maximum at 3730 cm^{-1}, characteristic for the stretching vibration of the O–H group, was missing in the samples used. The concentration of oxygen-containing impurities in the single crystal was ~$(2–5)\cdot 10^{16}$ cm^{-3}. Each batch of plates was irradiated under identical conditions; the irradiation time and consequently the irradiation dose were varied for different batches. It was not possible to determine the irradiation dose, since the plates were placed in front of the radiation source in a metal vessel filled with liquid nitrogen, i.e., the irradiation was carried out under non-standard conditions.

Three methods of lithium fluoride nanocrystals production have been used: films evaporation in vacuum; evaporation of the drops of LiF saturated water solution on the substrates; mechanical crushing of crystals up to the nanoparticles size with further pressing pellets of them. The majority of experiments has been carried out with nanocrystals produced by the third method. The sizes of certain nanocrystals amounted to hundreds nanometers. The obtained pellets were irradiated with γ-rays in the same way as the crystal plates.

After irradiation was finished, samples of crystalline plates or nanoparticles pellets from the same batch were successively removed from the liquid nitrogen.

The photoluminescence (PL) and optical density (last one only for plates) of the samples were measured for different stabilized temperatures $T > T_{\upsilon}$. The time required to go from the temperature T_{irrad} to the measurement (annealing) temperature T_{ann} was 1.5–2.0 min, so that no V_k centers or free electrons remained in the crystals at the beginning of the measurements. The annealing temperatures were varied in the range 283–328 K. At the lower temperatures in the crystalline plates it is impossible to receive during the period of experiment a database on the defects aggregation with participation of the F_2^+ centers sufficient for mathematical processing since diffusion of these centers occurs slowly. Annealing of the samples led to the formation of aggregate color centers. For each of the temperatures T_{ann} used, the time dependences of the PL intensity J were measured for several hours for F_2^+, F_2, $F_3(R_2)$ and F_3^+ color centers, respectively at wavelengths λ_{det} = 890, 680, 480 and 530 nm, and for excitation of PL by radiation at wavelengths of 630, 446, 380 and 420 nm. The data on PL spectra of $F_3(R_2)$ centers are taken from [21], and the data on all other PL and absorption spectra of the investigated CCs – from paper [4]. The PL excitation wavelengths are chosen so that to increase excitation selectivity. PL registration has been carried out on the wavelengths each of which ensured measurement of radiation of only one chosen CCs type.

The PL intensities were recorded on an SFL-1211 A spectrofluorimeter (SOLAR, Belarus) with signal averaging for 2 s. Since the optical densities at the excitation wavelengths were much less than unity (Figure 1) and also there was no reabsorption of photoluminescence, the PL intensities were proportional to the color center concentrations. On a Cary 500 Scan spectrophotometer (Varian, USA), we measured the optical densities D_1 at the maximum wavelength (λ = 248 nm) of the F absorption band of the F_1 centers (or at another wavelength within this band for larger values of D_1). The detected signal was averaged over 1 s.

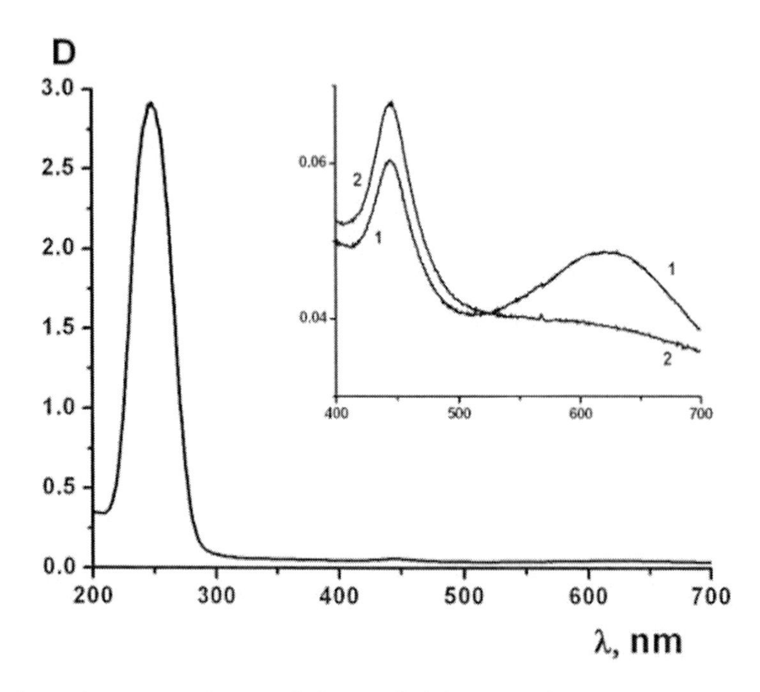

Figure 1. The absorption spectra of a crystal plate studied: 3.5 (1) and 24 hours (2) after the beginning of annealing at room temperature.

The recorded optical densities are proportional to the concentrations of the F_1 centers and were used for determination of these concentrations. The measured time dependences of the PL intensities represent the kinetics of variation in the concentrations of F_2^+, F_2, $F_3(R_2)$ and F_3^+ centers, and are the basis for analysis of color center aggregation processes in the post-radiation period. The temporal photoluminescence decay kinetics was measured by time-correlated photon counting. The samples were excited by pulses from a PLS450 light-emitting diode (PicoQuant, Germany).

III. Color Centers Aggregation Processes and Laws

So, the initial color center formation processes in the post-radiation period are determined by vacancies, F_1 and F_1^- centers, interstitial fluorine ions (I) and fluorine atoms (H). After the appearance of complex color centers, they also participate in subsequent transformations. In considering all the aggregation processes, we assume that only vacancies and F_2^+ centers migrate through the crystal. The observed long-term stability of the concentrations of F_1, F_2, F_3^+, F_3, F_4 color centers for more than a year after completion of the aggregation reactions allows us to conclude that we can neglect diffusion of these centers and also interstitial fluoride ions and fluorine atoms at the temperatures used in the experiments.

In the initial annealing period, the aggregation processes are due to the diffusion of vacancies created during irradiation. Then aggregate CCs can be formed as a result of the following reactions:

$$\upsilon_a + F_1 \rightarrow F_2^+, \qquad (1)$$

$$\upsilon_a + F_1^- \rightarrow F_2, \qquad (2)$$

$$\upsilon_a + F_2 \rightarrow F_3^+, \qquad (3)$$

$$\upsilon_a + I \rightarrow Fl^-, \qquad (4)$$

where Fl^- – fluorine ion in the lattice site. Reactions with participation of F_2^+ centers should occur in time intervals further removed from the beginning of the measurements (annealing of the crystal) than those with participation of vacancies created during irradiation:

$$F_2^+ + F_1 \rightarrow F_3^+, \qquad (5)$$

$$F_2^+ + F_1^- \rightarrow F_3, \qquad (6)$$

$$F_2^+ + H \rightarrow \upsilon_a + Fl^-, \qquad (7)$$

$$F_2^+ + I \rightarrow F_1 + Fl^-. \qquad (8)$$

Both vacancies created during irradiation of the crystal and vacancies formed as a result of reaction (7) participate in reactions (1) – (4). The former exist from the time that annealing of the sample begins; the latter arise some time after the beginning of annealing, when F_2^+ CCs are formed and then enter into reaction (7). Participation and the role of reaction (7) in post-radiation aggregation of color centers in LiF crystals, and its effect on the lifetime of F_2^+ centers, to our knowledge, have not been studied so far and need to be examined.

An example of the time dependence of the F_2^+ centers photoluminescence intensity (J(t)) during annealing of a crystal plate is shown in Figure 2 (curve 1). For all the irradiation doses and annealing temperatures, such dependences demonstrate similar behavior. In the initial period of time we observe a fast increase in the PL intensity (concentration) of the centers, in the last period of time we observe a slow decrease. Both the rise and fall in concentration are accelerated as the annealing temperature increases. Concentration growth is specified by development of reaction (1). Concentration falling occurs owing to reactions (5) – (8).

The time dependence of the F_2^+ centers photoluminescence intensity for pellets consisting of nanocrystals is presented by curve 2 in Figure 2. The dependences of such a kind are registered for all the irradiation doses and annealing temperatures. During the initial period of annealing fast falling of the PL intensity is observed. Falling proceeds approximately for the period of 14 s at T_{ann} = 298 K. It is accelerated with growth of annealing temperature and is equal to approximately 3 s at T_{ann} = 313 K. It is necessary to find out the reasons of this process.

At the next stage of annealing concentration of F_2^+ centers grows within the total measurement time interval (about 10 hours). The growth rate in nanocrystals is much less than in case of a crystal plate. The growth occurs as a result of reaction (1). There is no decrease of concentration of F_2^+ centers in nanocrystals.

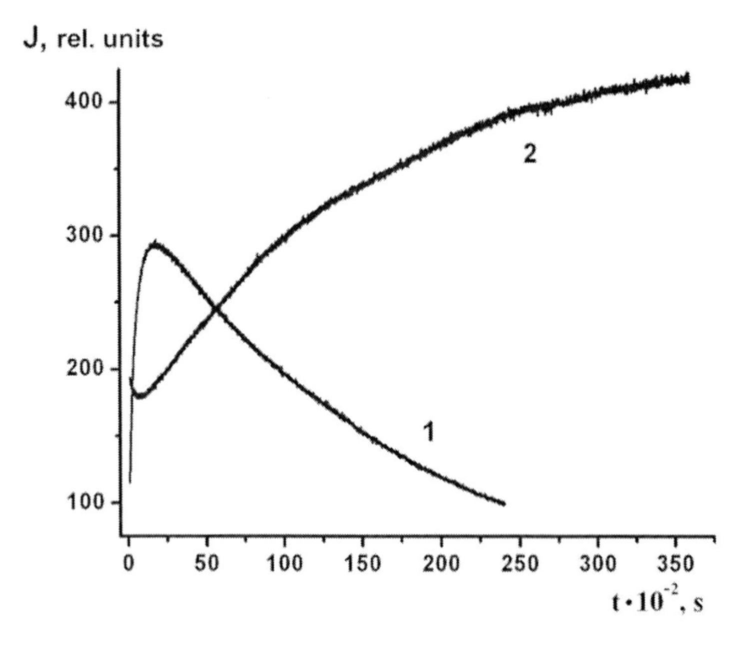

Figure 2. The time dependences of the photoluminescence intensities J for the F_2^+ centers in a crystal plate (1) and in a nanocrystalline pellet (2), annealing at T = 298 K.

The measurements show that the F_2^+ CCs remain stable: their concentration at a room temperature does is not being decreased within a year after irradiation.

The data on kinetics of changes of F_2^+ centers concentration are shown in Figure 3 in half-logarithmic scale. In order to display the concentration growth kinetics, the quantities $\ln[J_{fin} - J(t)]/J_{fin}$, where J_{fin} are the final PL intensities values at the stages, are used. For representation of kinetics of F_2^+ centers concentration decrease the quantities $\ln J(t)/J_{max}$, where J_{max} is the maximum value of the PL intensity, are used. The presented dependences both at the stage of growth and at the stage of falling of concentrations at all used irradiation doses and annealing temperatures are well described by straight lines.

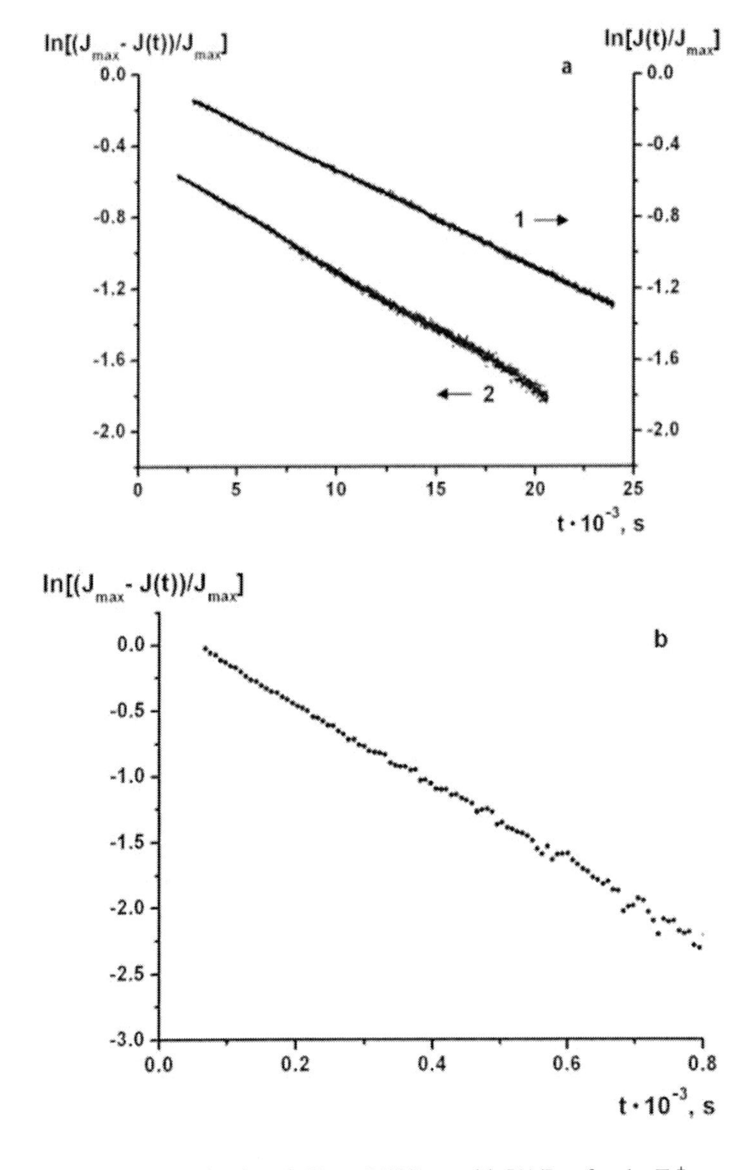

Figure 3. The time dependences of values $\ln[J_{fin} - J(t)]/J_{fin}$ and $\ln J(t)/J_{max}$ for the F_2^+ centers concentrations in a crystal plate (a1, b) and in a nanocrystals (a2), annealing at T = 298 K.

It testifies about exponential nature of growth or decrease of centers concentration N within the corresponding time intervals:

$$N = N_0\left[1-\exp\left(-\frac{t}{\tau}\right)\right], \tag{9}$$

$$N = N_0\exp\left(-\frac{t}{\tau}\right), \tag{10}$$

where N_0 is the final (in (9)) or initial (in (10)) concentration of the defects, which are being formed or destroyed as a result of reactions; τ is a reaction time constant. This time constant depends on annealing temperature and, as it is easy to show, is an average time of an overall reaction. The constants are being determined by the measured dependences J(t).

The F_2 and F_3^+ CCs concentrations in the post-radiation period grow during the whole annealing process both in crystal plates and in nanocrystals (Figure 4). The time required to complete the growth depends on the annealing temperature. At room temperature, it takes about 24 hours. In all the measurements with crystalline plates, the coefficient describing increase over the entire annealing time is higher for F_3^+ centers than for F_2 centers. For example, the data presented in Figure 4 show that concentration of F_3^+ centers increases by a factor of 6.7, while concentration of F_2 centers increases by a factor of 5.6. It follows also from the data presented in Figure 4 that for nanocrystals concentration of F_3^+ centers increases by a factor of 2.2, while concentration of F_2 centers increases by a factor of 3.3.

For curves 1, 2 in Figure 4 two growth stages are characteristic: fast (short) and slow (long). The F_2 and F_3^+ color centers formation is determined by processes with participation of vacancies and F_2^+ centers. For the processes with participation of the vacancies, created in the irradiation stage, time constants corresponding to the vacancy lifetimes τ_υ are typical. Processes with participation of F_2^+ centers are characterized by times $\tau_2 >> \tau_\upsilon$. Therefore in the kinetics of the concentrations of F_2 and F_3^+ color centers, we observe (Figure 4) an initial (fast) and a final (slow) component. Formation of F_2 and F_3^+ centers with participation of vacancies in the initial stage of the post-radiation period can occur only as a result of reactions (2) and (3), i.e., $\upsilon_a + F_1^- \rightarrow F_2$ and $\upsilon_a + F_2 \rightarrow F_3^+$. Accordingly, we must acknowledge the existence of F_1^- centers which are generated during irradiation of the crystal. Formation of these CCs in the post-radiation period is not possible due to the absence of free electrons in this period. In the final stage of the post-radiation period, F_2^+ centers and vacancies migrating through the crystal sample participate in aggregation processes. The vacancies are produced as a result of reaction (7) when F_2^+ centers encounter interstitial fluorine atoms. The vacancies formed in this way, participating in reactions (2) and (3), generate F_2 and F_3^+ color centers. So, the processes described for the formation of color centers during this stage are complicated. Each of them consists of two elementary reactions. Note that the F_3^+ centers in this stage can also be formed as a result of reaction (5). Thus the entire process for the formation of F_3^+ CCs is even more complicated.

Note that in our experiments, we did not obtain any data indicating the presence of electrons in the samples in the post-radiation period. At the first fast stage of concentrations growth at all used irradiation doses and annealing temperatures the dependences of values

$\ln[J_{fin} - J(t)]/J_{fin}$ for F_2 and F_3^+ centers in crystal plates are well described by straight lines (Figure 5). It confirms the exponential character of growth of these concentrations. It should be noted that tangents of inclination of lines to the abscissa axis in Figures 3 and 5 are equal to values τ^{-1} which are presented in equations 9 and 10.

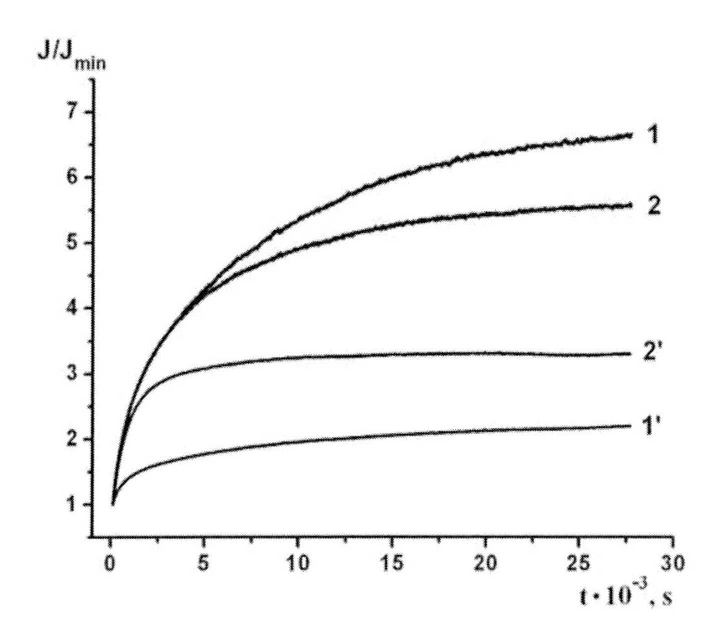

Figure 4. The time dependences of the photoluminescence intensities for the F_3^+ (1, 1′) and F_2 (2, 2′) centers in a crystal plate (1, 2) and in a nanocrystalline pellet (1′, 2′), annealing at T = 298 K.

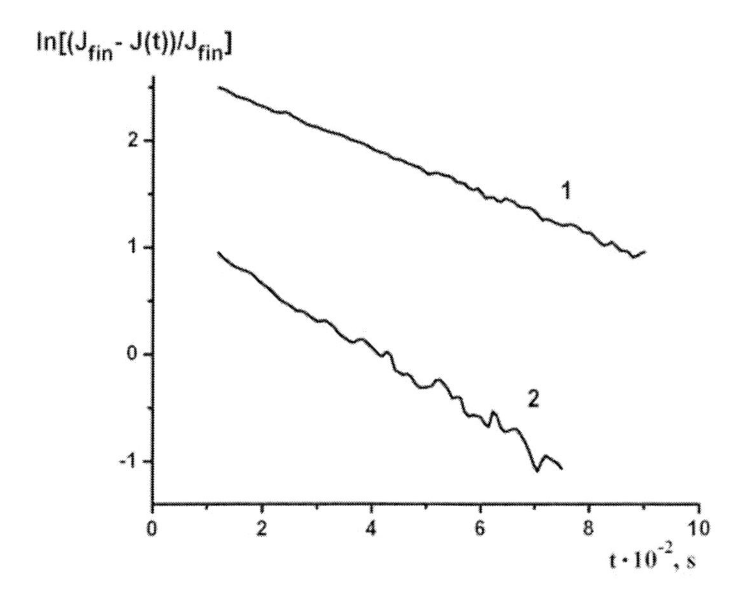

Figure 5. The time dependences of values ln[Jfin – J(t)]/Jfin in the initial fast stage of the F_2 (1) and F_3^+ (2) centers concentrations growth in a crystal plate.

For the fast initial exponential stage of rise in the concentrations of F_2^+, F_3^+ and F_2 centers, we obtained the following results. The time constants (exponential time constants) for the reaction (3) of formation of F_3^+ centers are smaller than the time constants for the process (reactions (2) and (3)) of formation of F_2 centers, measured at the same annealing temperatures in samples irradiated by the same dose. The time constants for both types of centers are greater than the life times $\tau_{1\upsilon}$ of the vacancies found under the same conditions from the photoluminescence kinetics of the F_2^+ centers.

Measurements of the F_3 centers PL kinetics at various annealing temperatures T_{ann} have been carried out only for one batch of the crystalline plates. The used conditions of irradiation ($T_{irrad} << T_\upsilon$) and annealing at $T > T_\upsilon$ lead to reduced concentrations of these CCs, which is confirmed by experiments and published data [22]. So, their consequent low PL intensity has presented certain difficulties for the mathematical processing of results.

Figure 6 shows the intensity of the $F_3(R_2)$ centers photoluminescence as a function of time. The kinetics of the F_3 centers concentration displays two distinct stages, the first one fast and short, and the second one very slow and long. The first stage is well described by an exponential behavior, and its time constant is approximately equal to the analogous constant times for F_2 and F_3^+ centers, but longer than for F_2^+ centers.

The reasons of the fast growth of F_3 centers concentration at the initial stage of the kinetics still remain unclear. The second slow stage of the kinetics is described by an exponential (see Figure 6, curve 2) with the long time constant. The time of the beginning of the slow stage is approximately equal to the time when the photoluminescence of F_2^+ centers achieves its maximum value (see Figure 2, curve 1).

In the second slow stage of the kinetics F_3 centers can be formed only by means of reaction (6) as a result of F_2^+ centers diffusion and their aggregation with F_1^- centers.

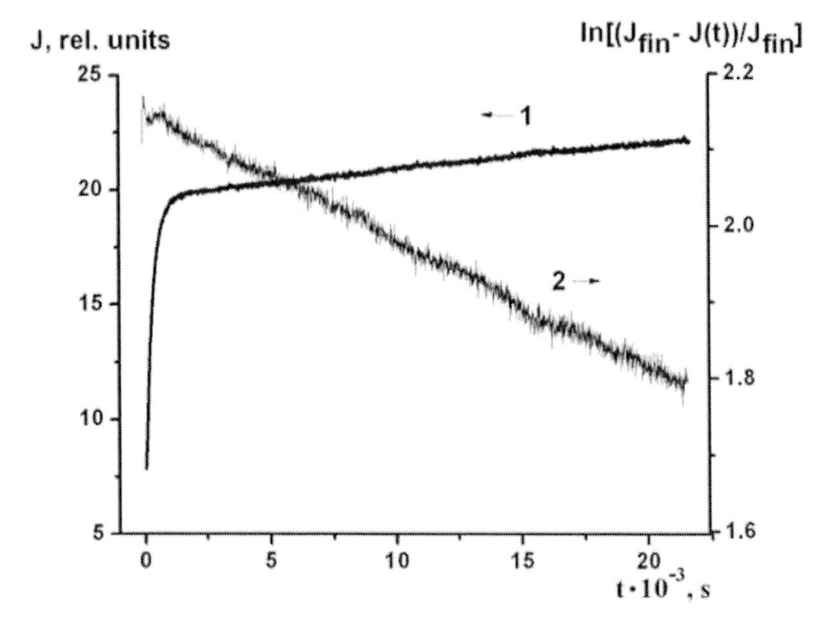

Figure 6. The time dependences of the PL intensity (1) and the value $\ln[J_{fin} - J(t)]/J_{fin}$ (2) for the F_3 centers in a crystal plate, annealing at T = 303 K.

It is now time to summarize the aggregation processes for F_2^+, F_2, F_3^+ and F_3 centers in the post-radiation period.

1. F_2^+ centers are formed as a result of the anion vacancies diffusion and the reaction $\upsilon_a + F_1 \rightarrow F_2^+$. The vacancies, created during the sample irradiation and as a consequence of reaction (7), are used for this formation. Thus, due to complex process consisting of reactions (7) and (1), reproduction of fresh F_2^+ centers is provided when $t >> \tau_\upsilon$. The F_2^+ centers life time is defined by reactions (5) – (8) and also by the reproduction of such centers with the process described by reactions (7) and (1).
2. In the post-radiation period, F_2 centers are formed with vacancies only as a result of the reaction $\upsilon_a + F^- \rightarrow F_2$. The vacancies, created during the irradiation and as a consequence of reaction (7), participate in the F_2 centers formation. The F_2 centers concentration decreases owing to the reaction $\upsilon_a + F_2 \rightarrow F_3^+$.
3. F_3^+ centers are formed as a result of the reaction $\upsilon_a + F_2 \rightarrow F_3^+$ in the initial stage of the annealing process and then as a consequence of both the reaction $F_2^+ + F_1 \rightarrow F_3^+$ and the reactions cascade: $F_2^+ + H \rightarrow \upsilon_a + Fl^-$ and $\upsilon_a + F_2 \rightarrow F_3^+$.
4. F_3 centers are formed owing to the reaction $F_2^+ + F_1^- \rightarrow F_3$. The formation processes in the first fast stage still remain unclear.

Kinetics of many reactions and processes of formation and destroy of radiation defects in bulk crystals and nanocrystals during the post-radiation period is well described by exponential dependences. It follows from the data presented in Figures 3, 5 and 6 that growth of concentrations of centers F_2^+, F_2, F_3^+ and F_3 as well as reduction of F_2^+ centers concentration is defined by formulas (9) and (10).

IV. Defects Diffusion

In the crystal plates under study there is no defects concentration gradient. Therefore defects diffusion in them takes place with equal probability in three mutually perpendicular directions. Such diffusion can be defined as random walk type diffusion. For it the following laws are typical [23]:

$$L^2 = 6Dt, \quad (11)$$

$$D = D_0 \exp\left(-\frac{E_a}{kT}\right), \quad (12)$$

where L^2 is a square of a defects diffusion path (square of displacement from the initial point) for time t, D is the diffusion coefficient, E_a is the activation energy of diffusion, k is the Boltzmann constant, T is a temperature of medium in which diffusion is observed, D_0 is a pre-exponential factor in the temperature dependence of the diffusion coefficient.

It follows from a relationship (11) that the aggregation reactions kinetics is being composed as a result of mobile defects diffusion. At the beginning (small values of times t) the mobile defects which have travelled the shortest diffusion pathway L enter into a reaction. Then in each next following point of time, the defects, which have travelled the longer diffusion path L as compared with the defects reacted at the previous moment, react.

The mathematical descriptions of the dependences of concentrations N on diffusion path L travelled by the mobile defect from the initial point to the meeting with the partner in a reaction, i.e. on distance L between the reacting defects, are derived from equations (9), (10) taking into account expression (11). They look like:

$$N = N_0 \left[1 - \exp\left(-\frac{L^2}{L_\tau^2} \right) \right]. \quad (13)$$

$$N = N_0 \exp\left(-\frac{L^2}{L_\tau^2} \right), \quad (14)$$

where L_τ^2 is the square of a diffusion path travelled by the defect for time τ. It follows from equations (13), (14) that the value L_τ^2 is equal to the mean value of diffusion pathways squares travelled by the mobile defects prior to their entering into a reaction for the whole period of reaction. Hence, this value is equal also to the mean value L_{ms}^2 of the squares of the distances between reacting partners.

The mean value of the squares of distances between reacting defects $L_{ms}^2 = L_\tau^2$ should depend on a crystal irradiation dose, but remain constant at annealing temperature variation. The confirmation of this position correctness should be found in the experiment results. Let us analyze these results. From equations (11), (12) at $t = \tau$ and $L^2 = L_\tau^2$ the following relationship is derived:

$$\ln\left(\frac{1}{\tau}\right) = \ln\left(\frac{6D_0}{L_\tau^2}\right) - \left(\frac{E_a}{k}\right)\left(\frac{1}{T}\right). \quad (15)$$

If the value L_τ^2 does not depend on ambient temperature, the experimentally obtained functions $\ln(1/\tau) = f(1/T)$ should be graphically represented by straight lines. Obtaining of straight lines will also testify that in the objects being studied by us the diffusion of random walks type takes place.

From the experimental time dependences J(t) of the PL intensities (similar shown in Figures 2, 4 and 6) for the F_2^+, F_2, F_3^+ and F_3 centers, we can determine the life times τ_υ and τ_2 of the vacancies and F_2^+ defects migrating through the samples and participating in formation of these centers. The time dependencies of the values $\ln[J_{fin} - J(t)]/J_{fin}$ and $\ln J(t)/J_{max}$ (see Figures 3, 5 and 6) demonstrate the possibilities of this determination. For samples from each batch irradiated by different doses we have determined the times τ_υ and τ_2 at several annealing temperatures. The determination procedure has been described earlier [9].

The temperature dependences of $\ln(1/\tau_\upsilon) = f(1/T)$ for vacancies are shown in Figure 7. The life times τ_υ were obtained from the functions J(t) measured for the F_2^+ centers. They characterize the reaction (1). The experimental data (open circles) are well fitted by straight lines. From crossing of the straight lines with an ordinate axis it is possible to find values

$6D_0/L_\tau^2$. The meaning of values $L_\tau^2 = L_{ms}^2$ obviously should be defined by the average concentration of reacting defects.

The tangents of the lines slopes relative to abscissas are equal to E_a/k and hence allow us to calculate the activation energies. According to five different experiments we have found the following value $E_{a\upsilon} = 0.60 \pm 0.02$ eV for vacancies in the crystalline plates, where the error is given by the root-mean-squares of the distribution. From the data 3, represented in Figure 7, we determine the activation energy $E^*_{a\upsilon} = 1.17 \pm 0.06$ eV for vacancies diffusion in the nanocrystals.

For the same annealing temperatures, the life times of vacancies are shorter in bulk crystals irradiated by higher doses (Figure 7, compare data 1 and 2). In particular, when increasing the irradiation dose, we have observed the life time decrease from 233 to 135 s at annealing temperature T = 295 K. The fall in the life times τ_υ is due to the increase in the concentrations of the vacancies and of the F_1 centers and to the reduction in the value $L_{ms}^2 = L_\tau^2$ as the irradiation dose increases. This value demagnification results in the more short life times according to the equation $L_\tau^2 = 6D\tau_\upsilon$.

The F_3^+ CCs concentrations growth takes place owing to reaction (3). The variation of the F_2 defects concentrations is defined by the process consisting of two elementary reactions (2) and (3).

At various annealing temperatures, the kinetics of the F_2 and F_3^+ defects formation has been measured in crystal plates irradiated with equal dose. The meaning of values $\ln(1/\tau_\upsilon)$ and 1/T have been determined from the obtained data for the initial stage of concentrations growth. These values are presented by open circles in Figure 8. It should be noted, that the mean times τ_υ for reactions (1), (2) and (3) are different.

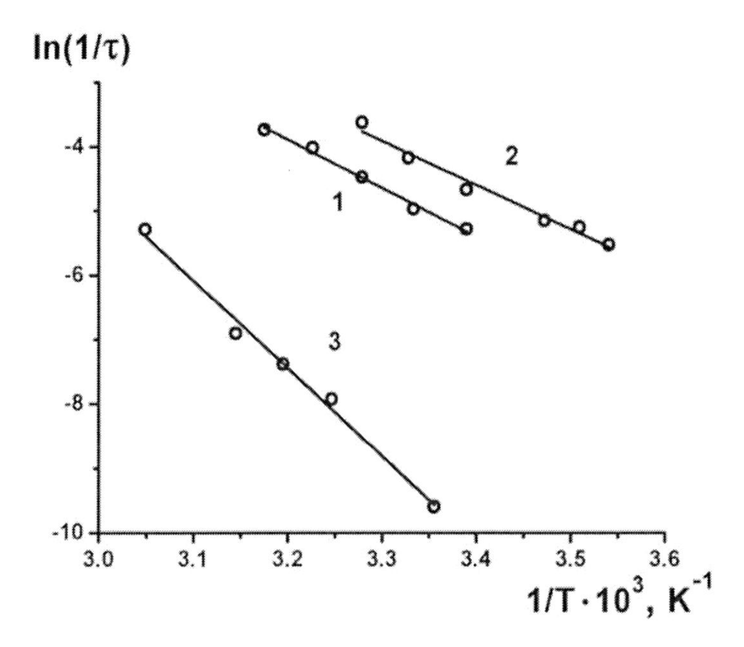

Figure 7. The dependences $\ln(1/\tau_\upsilon) = f(1/T)$ for vacancies (the reaction $\upsilon_a + F_1 \rightarrow F_2^+$) in the crystalline plates (1, 2) at irradiation doses increasing in the sequence 1 → 2 and in the pellets with nanocrystals (3).

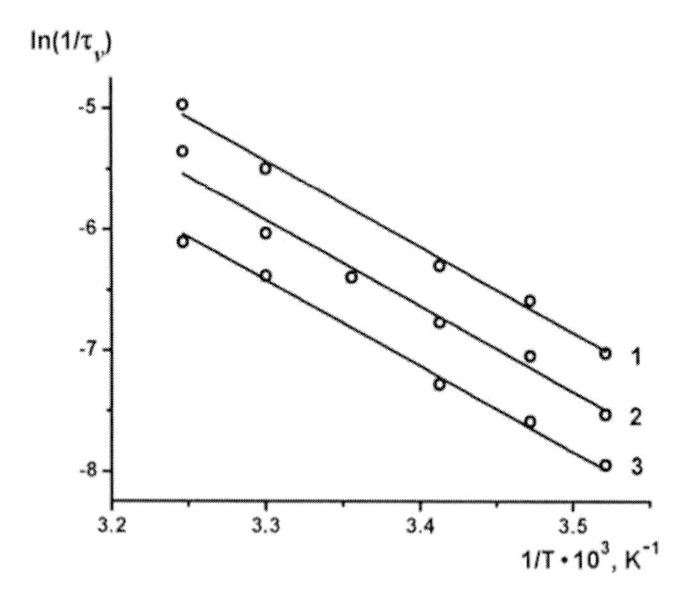

Figure 8. The dependences $\ln(1/\tau_\upsilon) = f(1/T)$ for reactions $\upsilon_a + F_1^- \rightarrow F_2$ (1), $\upsilon_a + F_2 \rightarrow F_3^+$ (2) and for the process consisting of the reactions $\upsilon_a + F_1^- \rightarrow F_2$ и $\upsilon_a + F_2 \rightarrow F_3^+$ (3) for crystal plates irradiated with the same dose. The experimental results (2, 3) and the results (1) of calculation in accordance with the formula (16) are shown by open circles.

Experimental data (Figure 8) are well approximated by straight lines with slopes corresponding to an activation energy of vacancies diffusion $E_{a\upsilon} = 0.60$ eV. This result can be predicted because in reactions (1) – (3) the mobile component is presented by vacancies. The mean times $\tau_{\upsilon 1}$ and $\tau_{\upsilon 3}$ of reactions (1) and (3) define rates $V_{\upsilon 1,2} = (\tau_{\upsilon 1,2})^{-1}$ of formations of F_2^+ and F_3^+ CCs at the initial stage of annealing. Rate $V'_{\upsilon 2}$ of formation of F_2 centers as a result of the process consisting of reactions (2) and (3) can be presented as follows:

$$V'_{\upsilon 2} = V'_{\upsilon 1} - V_{\upsilon 2}, \tag{16}$$

where $V'_{\upsilon 1}$ is a rate of reaction (2). Using relationships (16), (11), (12) it is possible to show, that the process consisting of two reactions (2), (3) and defining a variation of the F_2 CCs concentrations is described by the values of activation energy $E_{a\upsilon}$ and parameter $D_{0\upsilon}$ which are characteristic for vacancies. The equality (16) allows to find the mean times $\tau'_{\upsilon 1} = 1/V'_{\upsilon 1}$ for reaction (2). The dependence (15) for the calculated times $\tau'_{\upsilon 1}$ is presented by the straight line 1 in Figure 8. From the received data the hierarchy of reactions (1), (2), (3) by their rates follows.

Reaction (1) during which defects F_2^+ are being formed run more rapidly than other reactions. Centers F_2 are being formed with smaller velocity owing to reaction (2). Finally, centers F_3^+ are being formed, as the product of reaction (3), most slowly. This reduction of velocities indicates the decrease of frequency of vacancies meetings with partners F_1, F_1^- and F_2 in reactions (1), (2) and (3) accordingly. Therefore it is possible to conclude, that the values of concentrations N_1, N_1' and N_2 for the F_1, F_1^- and F_2 centers accordingly satisfy the following inequality: $N_1 > N_1' > N_2$.

The decreasing stage of the time dependence of the F_2^+ centers concentration (Figure 2, curve 1) is described by a time constant which corresponds to the life time τ_2 of these centers.

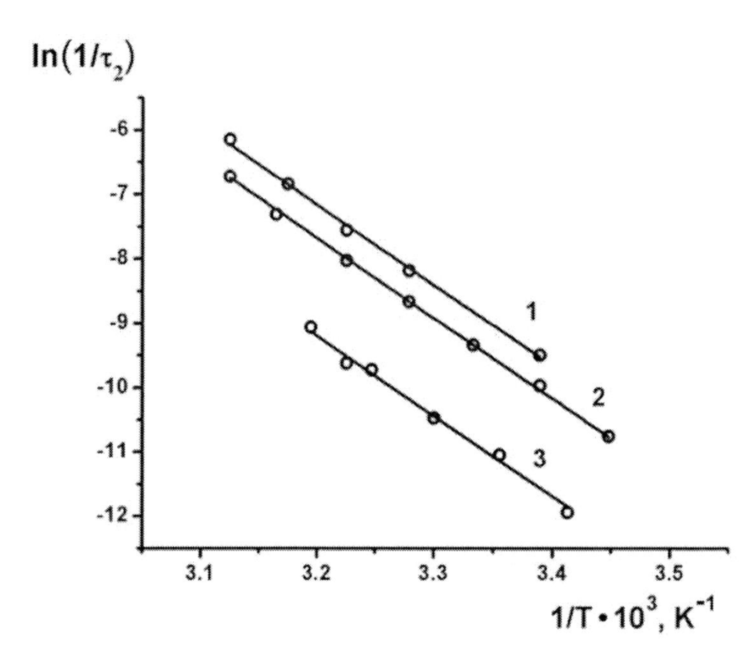

Figure 9. The dependences $\ln(1/\tau_2) = f(1/T)$ for the F_2^+ centers concentration decrease at various radiation doses (1, 2) and for the $F_3(R_2)$ centers concentration growth (3) in crystal plates. Radiation doses are increased in sequence 1→2. The experimental data are shown by open circles; the straight lines are approximations of the experimental data.

The life times τ_2 have been determined at the different irradiation doses for crystal plates. When increasing the irradiation dose, the life time τ_2 increases from $13.3 \cdot 10^3$ to $21.5 \cdot 10^3$ s at annealing temperature $T_{ann} = 295$ K. So, the ratio τ_2/τ_υ is approximately 10^2. The difference in the values of parameters τ_2 and τ_υ is caused, first of all, by the difference of diffusion coefficients values for vacancies (D_υ) and F_2^+ centers (D_2). The dependences $\ln(1/\tau_2) = f(1/T)$, characterizing the F_2^+ centers diffusion, are shown in Figure 9. The experimental data are well fitted by straight lines, the slopes of which allow us to determine the activation energy for the F_2^+ centers diffusion. According to six experiments, we have found the following value: $E_{a2} = 1.07 \pm 0.01$ eV.

To understand the reasons of value τ_2 variation with the radiation dose growth we will examine reactions (5) – (8) in which centers F_2^+ participate at times $t >> \tau_\upsilon$. As a result of processes (5), (6) and (8) the stable formations are created, and the F_2^+ centers irrevocably disappear. Reaction (7) leads to anionic vacancies formation. The appeared vacancies participate in transformations (1) – (4). The special interest represents the fact, that the F_2^+ centers are being reproduced as a result of the reaction (1). We will show, that exactly by reproduction of these centers due to the process consisting of reactions (7) and (1) it is possible to explain growth of time τ_2 with crystal irradiation dose increase.

A rate $V_2 = 1/\tau_2$ for decrease of F_2 CCs concentrations can be written as follows:

$$V_2 = V_{21} + V_{21}^{/} + \frac{V_{2H}(V_{\upsilon 1}^{/} + V_{\upsilon 2})}{(V_{\upsilon 1} + V_{\upsilon 1}^{/} + V_{\upsilon 2})}, \tag{17}$$

where V_{21}, V'_{21}, and V_{2H} are the rates of reactions (5), (6) and (7) accordingly. The rates $V_{\upsilon 1}$, $V'_{\upsilon 1}$ and $V_{\upsilon 2}$ characterize reactions (1), (2) and (3) accordingly, in which the vacancies, created as the result of reaction (7), participate. Increase of times $\tau_2 = V_2^{-1}$ with the radiation dose growth means that decrease of the last summand in (17) at this growth exceeds increase of the first two summands. Such situation can occur only at the expense of the fast increase of a rate $V_{\upsilon 1}$.

Equality (17) also indicates that the process of F_2^+ CCs concentration decrease in spite of its complexity is characterized by diffusion activation energy for these centers. The coordinates of crossings of dependences $\ln(1/\tau_2) = f(1/T)$ (Figure 9, straight lines 1, 2) with an ordinate axis are being defined by the value of the pre-exponential factor D_{02} and the values of the complicated combinations on the basis of distances between the defects participating in reactions (5) – (8) and (1) – (4).

The slow and long stage of the $F_3(R_2)$ centers concentration kinetics (Figure 6, curve 1) in crystalline plates is due to just one reaction (6) as a result of F_2^+ centers diffusion and their aggregation with F_1^- centers. This stage is described by an exponential (see Figure 6, straight line 2). The time constant of the exponential is longer than the F_2^+ centers lifetime τ_2. The time constants measured are reported in the usual configuration $\ln(1/\tau) = f(1/T)$ in Figure 9, curve 3.

The experimental data are well approximated by the straight line from which it is possible to retrieve the activation energy. The value $E_a = 1.08 \pm 0.03$ eV has been obtained for the activation energy, which coincides with the analogous quantity E_{a2} for F_2^+ centers diffusion, and testifies that F_3 centers at this final stage are being formed as a result of reaction (6). The experimental reality of reaction (6) is an additional proof of the existence of F_1^- centers in the crystal after irradiation.

V. The Distribution of Diffusion Parthways Travelled by the Mobile Defects Prior to Entry into Reactions

Thus, experimental studies show that the kinetics of reactions in the post-radiation period very often is well presented by the exponential law. A question arises which in this case should be distribution of the diffusion paths L travelled by the mobile defects prior to their entry into reaction, or, in other words, what is the probability of detection of the mobile defect, the diffusion path of which prior to entry into reaction belongs to the interval L ÷ L + dL.

From relationships (13), (14) the equation follows:

$$\frac{dN}{N_0 dL} = 2\frac{L}{L_{ms}^2}\exp\left(-\frac{L^2}{L_{ms}^2}\right) = Y(L). \qquad (18)$$

It defines a relative number of mobile defects which have travelled the diffusion path laying within the interval L ÷ L + dL before entry into a reaction with their localized partners and further disappearance as a result of this reaction. Hence, it also defines relative numbers of reacting partners, the distances between which lay within the same interval.

Distribution (18) is presented as a continuous function of distance L. However the defects, for example, vacancies, move by jumps. Each jump has certain length. Reacting defects are distributed by distances with a step which is approximately equal to a length of the elementary diffusion jump. These conditions can affect the distribution Y(L) only within the range of small distances L close to the length of this jump.

Function Y(L) satisfies the normalizing condition, i.e.

$$\int_0^\infty Y(L)\,dL = 1.$$

Integration within the range from zero to infinity does not mean that there are defects which travel very large diffusion pathways. The diffusion pathway should be also more than the interatomic distance in a crystal. Usage of the indicated integration limits should be considered only as the standard practice in physics (as, for example, at the analysis of the gas molecules distribution throughout velocities). Later we will show that the defects with very small or very large distances between them constitute an insignificant part of the total amount of defects. They make the insignificant contribution to the integral which allows to use such limits of integration.

The maximum of function (18) corresponds to the most probable distance L_m between reacting partners. Distance L_m is connected with a value L_{ms} as follows:

$$L_m = 2^{-0.5} L_{ms}. \tag{19}$$

The mean distance L_{av} between reacting partners can be found by a common rule of determination of the mean value which leads to the following equality:

$$L_{av} = \frac{\sqrt{\pi}\, L_{ms}}{2}. \tag{20}$$

Application of the same common rule for mean squared distances results, as one could expect, a value L_{ms}^2. It should be noted that $L_{ms} > L_{av} > L_m$ and L_{ms}: L_{av}: $L_m \approx 1.41$: 1.25: 1. The ratios between these values do not depend on annealing temperature of a crystal or irradiation dose.

The form of function Y(L) is presented in Figure 10 for two values of mean square root L_{ms}. These values are close to the values implemented in the experiments being carried out. With the increase of the mean squared distance the function maximum is being displaced, its value decreases and distribution is being widened. The areas under distributions remain equal to the unity irrespective of magnitude L_{ms}.

With using function (18) it is possible to calculate the probability with which the distances between components of reacting defects pairs are in certain intervals. For example, probabilities of detection of pairs with distances within intervals $L < L_{av}$ and $L > L_{av}$ are equal to 0.544 and 0.456 respectively. A share of pairs with the distances between components within intervals $L > 2.5L_{av}$ and $L < 0.1L_{av}$ are equal to 0.7 % and 0.78 % respectively. As it is possible to understand from the given numbers, the defects with very small and very large distances between them compose an insignificant part from the total number of the defects. They make the insignificant contribution to the integral written above. This circumstance serves as the ground for the the integration limits used above.

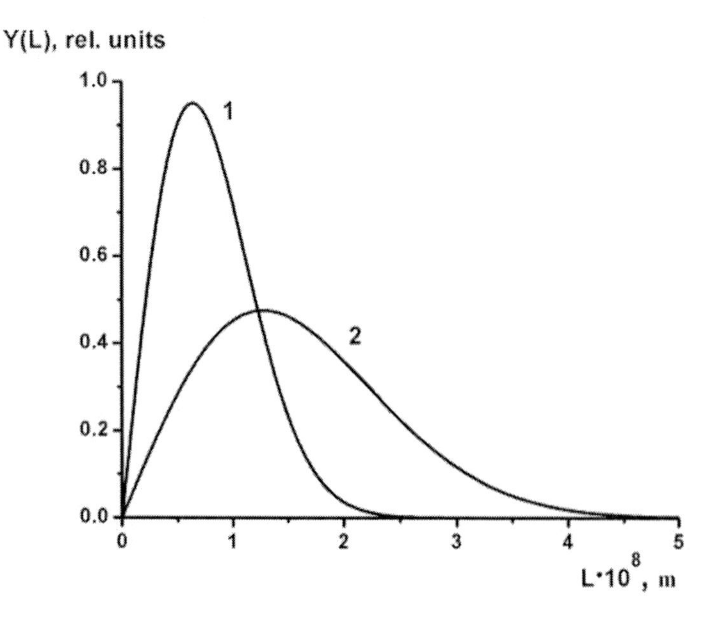

Figure 10. The functions of the reacting defects distribution throughout the distances between them at values $L_{ms} = 0.9 \cdot 10^{-8}$ (1) и $1.8 \cdot 10^{-8}$ m (2).

At derivation of the function describing the distribution of reacting defects by the distances between them no assumptions have been made. The foundation of its derivation is the experimentally established fact: the exponential character of the kinetics of the reaction products concentrations variation.

VI. The Diffusion Parameters for Mobile Defects in LiF Crystals

The relation (12) allows to determine the mobile defects diffusion coefficients if the values of pre-exponential factor D_0 and activation energy E_a of diffusion are known. The activation energies for diffusion of the vacancies and F_2^+ defects in crystal plates are already given in Section IV. For example, to calculate factor $D_{0\upsilon}$ for vacancies, it is necessary in the same experimental conditions to determine quantities $6D_{0\upsilon}/L_{ms}^2$ and values L_{ms}^2. Values $6D_{0\upsilon}/L_{ms}^2$ are being defined from the data presented in Figure 7 as it has been discussed considered in Section IV. For calculation of values L_{ms}^2 it is possible to use the data on concentrations N of the reacting defects.

Before determination of the absolute values of concentrations for various types of defects it is necessary to find the ratios between them. For this purpose we will use the data on dependences $\ln(1/\tau_\upsilon) = f(1/T)$ describing the kinetics of reactions (1), (2) and (3), as is shown by straight lines in Figures 7 and 8. The similar straight lines were received for samples irradiated with the same dose. Intersections of these straight lines with ordinate axis determine the values $6D_{0\upsilon}/L_{ims}^2$, where L_{ims}^2 with i = 1, 2, 3 are the average values of quadrates of diffusion pathways which be travelled by the vacancies prior to entry into reactions (1), (2), (3) respectively. The obtained values $6D_{0\upsilon}/L_{ims}^2$ give the ratios L_{1ms}^2: L_{2ms}^2: L_{3ms}^2.

The ratios L_{1ms}^2: L_{2ms}^2: L_{3ms}^2 = 1: 2.25: 3.6 have been found from the experimental data, and, as it follows from (20), we have L_{1av}: L_{2av}: L_{3av} = 1: 1.5: 1.9. The distinctions of diffusion paths values L_i are the consequence of distinctions in concentrations of the defects which participate in reactions (1), (2) and (3). The presented data were obtained for the samples with an irradiation dose at which the absorption coefficient k_{01} at the F-band maximum (F_l centers absorption) was equal to 56.8 cm^{-1}. It is possible to assume that $L_{av} \approx N^{-1/3}$. Then we obtain $N^*_1{}^{-1/3}$: $N^*_2{}^{-1/3}$: $N^*_3{}^{-1/3}$ = 1: 1.5: 1.9 where indices 1, 2 and 3 refer to concentrations N^* of the defects participating in reactions (1), (2) and (3), respectively.

The centers F_l and vacancies are involved in reaction (1). Usually the concentration of centers F_l is much more than the total concentration of other CCs types in the irradiated LiF crystals [22]. Thus the inequality $N_l >> N_\upsilon$ is valid, where N_l, N_υ are the concentrations of centers F_l and vacancies respectively and it is possible to suppose that $N^*_1 \approx N_l$ and $L_{1av} \approx N_l^{-1/3}$. The inequality $N_\upsilon >> N_2$ should be fulfilled for reaction (3) at the initial aggregation stage, because F_2 centers should be absent or present at small concentrations N_2 in the samples right after their extraction from the liquid nitrogen. Therefore it is possible to assume that $N^*_3 \approx N_\upsilon$ and the mean diffusion pathway L_{3av}, travelled by the vacancies prior to entry into reaction (3), is approximately equal to the mean distance between vacancies, i.e. $L_{3av} \approx N_\upsilon^{-1/3}$. Now it is possible to establish the ratio N_l: N^*_2: N_υ = 1: 0.30: 0.15.

Concentrations N_l of F_l centers can be determined from the results of measurement of their absorption coefficients using Smakula formula for Gauss profile of absorption [24]:

$$N_l f = 0.87 \cdot 10^{17} \frac{k_{01}\Delta n}{\left(n^2+2\right)^2}, \tag{21}$$

where f is the oscillator strength for absorption; Δ (in eV) is the width of the absorption band at half-height; n is the refractive index of the material at the wavelength of the absorption band maximum; k_{01} is the absorption coefficient (in cm^{-1}) at the same wavelength. The concentration N_l at the values f = 0.56, Δ = 0.76 eV, n = 1.42 taken from the literature and the measured value k_{01} = 56.8 sm^{-1} is equal to $5.9 \cdot 10^{17}$ cm^{-3}. Hence, for vacancies concentration we have $N_\upsilon \approx 8.9 \cdot 10^{16}$ cm^{-3} and for concentration of F_l^- centers and the vacancies participating in reaction (2) we obtain $N^*_2 \approx 1.8 \cdot 10^{17}$ cm^{-3}. The value N^*_2 is approximately twice more than the vacancies concentration. Therefore it is possible to suppose that the concentrations of both F_l^- centers and vacancies are approximately equal, i.e. $N_\upsilon \approx N_l'$.

The condition of electrical neutrality of a crystal for anionic component at the beginning of annealing process is of follows:

$$N_\upsilon = N_l' + N_I, \tag{22}$$

where N_I is the concentration of interstitial fluorine ions. Taking into account that $N_\upsilon \approx N_l'$ it is possible to assume the insignificant concentration of interstitial fluorine ions during the post-radiation period in LiF crystals gamma irradiated at low temperatures. Therefore reactions (4) and (8) cannot be considered at examination of the post-radiation transformation of the defects in LiF crystals.

The values $6D_{0\upsilon}/L_{1ms}^{2}$ and concentrations of F_1 centers have been determined for the samples irradiated with various doses. Using the obtained data the value of pre-exponential factor $D_{0\upsilon} \approx 3 \cdot 10^{-9}\, m^2 s^{-1}$ for vacancies has been found. The basic error in the found value $D_{0\upsilon}$ is brought by assumption that $L_{1av} \approx N_1^{-1/3}$. In connection with these assumptions the error of the definition can be approximately 100%. Knowledge of the factor $D_{0\upsilon}$ and the diffusion activation energy $E_{a\upsilon}$ allows to calculate the diffusion coefficients D_υ for anionic vacancies in lithium fluoride crystals at various temperatures. For example, at temperature T = 303 K we receive the value $D_\upsilon \approx 3 \cdot 10^{-19}\, m^2 s^{-1}$ for the vacancies diffusion coefficient.

Centers F_3 are formed as a result of reaction (6) during the post-radiation period at the annealing stage when $t > \tau_\upsilon$. The kinetics of their formation is being defined by diffusion characteristics of F_2^+ centers and the mean distance between the reacting defects. Crossing of straight line 3 (Figure 9) with an ordinate axis gives the value $\ln(6D_{02}/L_{ms}^{2}) = 30.6$, where D_{02} is the pre-exponential factor in the equation (12) for F_2^+ defects, L_{ms}^{2} is the mean square of the diffusion pathways travelled by F_2^+ defects prior to their entry into reaction (6). As well as earlier at consideration of the vacancies diffusion, we will evaluate the mean square L_{ms}^{2} on the ground of value $L_{av} \approx N^{*-1/3}$, where N^* is the sum of the concentrations of F_2^+ and F_1^- defects participating in reaction (6). Taking into account the previously described ratios, we can suppose that $N^* \approx 0.3N_1$. Concentration N_1 of F_1 centers can be calculated by the measured absorption coefficient $k_{01} = 28.4\ cm^{-1}$ using relationship (21). Considering all above stated we find $D_{02} \approx 2 \cdot 10^{-3}\ m^2 s^{-1}$. Since the diffusion activation energy for F_2^+ centers is equal to 1.07 eV we obtain the value $D_2 \approx 3 \cdot 10^{-21}\ m^2 s^{-1}$ for the diffusion coefficient at temperature T = 303 K. It is two orders less than the diffusion coefficient for anion vacancies.

VII. The Features of the Defects Aggregation in LiF Nanocrystals

In this section let us summarize the obtained results on the radiation-induced defects diffusion and aggregation in lithium fluoride nanocrystals. Unfortunately, the tablets with nanocrystals being used in our experiments were not transparent. This circumstance did not allow to measure the absorption spectra of nanocrystals and to compare them with the spectra of crystals.

1. Distinctions between the luminescence spectra of nanocrystals and those of crystal plates have not been detected. It is necessary to study in more detail the luminescence spectra of nanocrystals.
2. The diffusion activation energy for anionic vacancies in nanocrystals is equal to 1.17 eV while for crystals it is 0.60 eV. In view of this the aggregation processes with participation of vacancies run much more slowly in nanocrystals as compared with bulk crystals.
3. The diffusion activation energy for F_2^+ centers in nanocrystals is so high that their diffusion was not detected at used annealing temperatures. Therefore in the post-radiation period decrease of these centers concentration was not observed. Moreover,

within a year reduction of their photoluminescence intensity under stable excitation conditions was not detected. Thus, in nanocrystals reactions (5) – (8) do not take place, and F_2^+ centers do not participate in the aggregation processes during the post-radiation period at temperatures close to the room temperature.

4. The carried out luminescence decay time measurements for F_2^+ and F_3^+ centers in nanocrystals give the values 20.9 and 8.3 ns respectively. The luminescence decay time kinetics is well represented by one exponential curve. These results correspond to the literature data for the crystals [25, 21]. Taking into account coincidence of photoluminescence spectra and luminescence decay times in bulk crystals and nanocrystals for F_2^+ and F_3^+ centers it is possible to conclude with high reliability that these CCs in both types of samples have the identical structure and electronic properties.
5. Two exponents are required for mathematical simulation of experimentally measured luminescence decay curves of F_2 centers (λ_{det} = 670 nm) in nanocrystals. The time constants of these exponential curves are equal to 4.6 and 16.0 ns. The last value coincides with the luminescence decay time of F_2 centers in bulk crystals [21]. Presence of a component with the unknown earlier time constant indicates appearance in nanocrystals of the new type of defects which are not detected in bulk crystals. The photoluminescence band of these defects overlaps with the photoluminescence spectrum of F_2 centers. Additional investigations are required for the final decision of this problem.
6. The characteristic times determined from the measured kinetics of growth of F_2 and F_3^+ CCs concentrations in nanocrystals at various annealing temperatures in the post-radiation period do not form straight-line dependence $\ln(1/\tau_\upsilon) = f(1/T)$. This result is partly in agreement with the conclusion of the previous paragraph about appearance of the new type of defects in nanocrystals.

CONCLUSION

The investigated kinetics of the reactions in which the intrinsic defects participate is described in many cases by the exponential law. On the basis of such law and the fact, that in the investigated cases the random walk diffusion occurs, a distribution of the diffusion pathways travelled by the mobile defects prior to entry into a reaction has been found. Distribution of diffusion pathways by their values corresponds to distribution of the reacting defects by the distances between them.

In the carried out investigations in lithium fluoride nanocrystals during the post-radiation period only vacancies migration has been detected. This migration is responsible for the post-radiation color centers aggregation. The vacancies diffusion coefficient in nanocrystals is twice less than that in crystals. In this connection the processes with participation of the vacancies in crystals run faster than in nanocrystals. For example, it follows from the data of Figure 2 that the time constant characterizing duration of F_2^+ centers formation process is equal to $2.3 \cdot 10^2$ s for crystals and $6 \cdot 10^3$ s for nanocrystals.

Migration of F_2^+ centers in nanocrystals has not been observed. Moreover, reduction of their concentration has not been detected after termination of annealing for the long time observations at room temperature. Therefore these centers are stabilized. The issue of participation of the unstabilized F_2^+ centers in the aggregation processes in nanocrystals remains unclarified. It is also unknown share of the unstabilized centers in the total concentration of F_2^+ centers and how does this share depend on crystals sizes.

We have to define the type as well as the spectral characteristics of the color centers with small photoluminescence decay time (4.6 ns). Such CCs type is absent or, at least, it is not being detected in bulk crystals.

The last but not least definition of the reasons determining the differences of radiation defects formation in nanocrystals as compared with bulk crystals demands additional study and conducting of new experiments.

Acknowledgments

The author would like to thank V. S. Kalinov, A. P. Stupak and L. P. Runets for their general involvement in this subject and useful discussions.

References

[1] Kozlovski, V.; Abrosimova, V. Radiation Defect Engineering (Selected Topics in Electronics and Systems); *World Scientific Pub. Co.: Singapore,* 2005; pp 1–253.

[2] Furetta, C.; Weng, P.-S. Operational Thermoluminescence Dosimetry; *World Scientific Publ. Co.: Singapore,* 1998; pp 1–255.

[3] Basiev, T. T.; Zverev, P. G.; Mirov, S. B. In *Handbook of Laser Technology and Applications;* Webb, C. E.; Jones J. D. C.; Eds.; Boca Raton, Taylor and Francis Group, CRC Press: Philadelphia, USA, 2003; pp 499–522.

[4] Montereali, R. M. In *Handbook of Thin Film Materials*; Nalwa, H.S.; Ed.; Academic Press: New York, USA, 2002; Vol. 3, pp 399–431.

[5] Kurobori, T.; Kawamura, K.; Hirano, M.; Hosono, H. *J. Phys.: Condens. Matter.* 2003, vol. 15, L399–L405.

[6] Voitovich, A. P.; Kalinov, V. S.; Loiko, Yu. V.; Naumenko, N. N.; Runets, L. P.; Stupak, A. P. *J. Appl. Spectr.* 2008, vol. 75, 104–113.

[7] Lisitsyna, L. A. *Fiz. Tverd. Tela.* 1992, vol. 34, 2694–2705.

[8] Watkins, G.D. *J. Appl. Phys.* 2008, vol. 103, 106106.

[9] Voitovich, A. P.; Voitikova, M. V.; Kalinov, V. S.; Martynovich, E. F.; Novikov, A. N.; Runets, L. P.; Stupak, A. P.; Montereali, R. M.; Baldacchini, G. *J. Appl. Spectr.* 2011, vol. 77, 857–868.

[10] Makarenko, L. F.; Korshunov, F. P.; Lastovski, S. B.; Murin L. I.; Moll, M. *Dokl. Nat. Akad. Nauk Belarus'*. 2011, vol. 55, N 3, 49–54.

[11] McLaughlin, W. L.; Miller, A.; Ellis, S. C.; Lucas, A. C.; Kapsar, B. M. *Nucl. Instr. Methods Phys. Res.* 1980, vol. 175, 17–18.

[12] Lakshmanan, A. R.; Madhusoodanan, U.; Natarajan, A.; Panigrahy, B. S.. *Phys. Status Solidi* (a). 1996, vol. 153, 265–273.

[13] Kovacs, A.; Baranyai, M.; McLaughlin, W. L.; Miller, S. D.; Miller, A.; Fuochi, P. G.; Lavalle, M.; Slezsak; I. *Rad. Phys. Chem.* 2000, vol. 57, 691–695.

[14] Almaviva, S.; Marinelli, M.; Milani, E.; Prestopino, G.; Tucciarone, A.; Verona-Rinati, C.; Angelone, M.; Lattanzi, D.; Pillon, M.; Montereali, R. M.; Vincenti, M. A. *J. Appl. Phys.* 2008, vol. 103, 054501.

[15] Basiev, T. T.; Voronov, V. V.; Konyushkin, V. A.; Kuznetsov, S. V.; Lavrishchev, S. V.; Osiko, V. V.; Fedorov, P. P.; Ankudinov, A. B.; Alymov, M. I. *Doklady Physics.* 2007, vol. 417, 635–638.

[16] Nahum, J.; Wiegand, D.A. *Phys. Rev.* 1967, vol. 154, 817–830.

[17] Nahum, *J. Phys. Rev.* 1967, vol. 158, 814–825.

[18] Gu, H.; Qi, L.; Wan, L.; Guo, H. Opt. Commun. 1989, vol. 70, 141–144.

[19] Baryshnikov, V. I.; Kolesnikova, T. A.; Dorokhov, S. V. *Opt. Spektrosk.* 2000, vol. 89, 70–75.

[20] Voitovich, A. P.; Kalinov, V. S.; Naumenko, N. N.; Runets, L. P.; Stupak, A. P. *J. Appl. Spectr.* 2010, vol. 77, 247–254.

[21] Voitovich, A. P.; Kalinov, V. S.; Naumenko, N. N.; Stupak, A. P. J. *Appl. Spectr.* 2006, vol. 73, 866–873.

[22] Baldacchini, G.; Montereali, R. M.; Nichelatti, E.; Kalinov, V. S.; Voitovich, A. P.; Davidson, A. T.; Kozakiewicz, A. G. *J. Appl. Phys.* 2008, vol. 104, 063712, 1–10.

[23] Frenkel, J. *Kinetic Theory of Liquids*; Dover Public.: New York, USA, 1955; pp 1– 488.

[24] Agullo-Lopez, F.; Catlow, C. R. A.; Townsend, P. D. *Point Defects in Materials*; Academic Press Inc.: San Diego, US, 1988; pp 1–445.

[25] Mirov, S. B.; Dergachev, A. Yu. In Solid States Lasers; Scheps, R.; Ed.; *Proceed. SPIE;* 1997; Vol. 2896, pp 162–173.

In: Ionizing Radiation ISBN: 978-1-62257-343-1
Editors: Eduard Belotserkovsky and Ziven Ostaltsov

Chapter 4

MECHANICAL AND OPTICAL PROPERTIES OF IONIC CRYSTALS, EXPOSED TO THE COMBINED ACTION OF VARIOUS EXTERNAL FIELDS (*REVIEW*)

***V. Kvatchadze*[1] *and M. Galustashvili*[2]**
I. Javakhishvili Tbilisi State University,
E. Andronikashvili Institute of Physics,
Tbilisi, Georgia

ABSTRACT

The review covers the main body of works carried out from 1975 to 2012 years. They are unified by the general idea – to study experimentally effect of ionizing radiation combined with various external fields (mechanical stress, electrical and magnetic fields etc.) upon optical-mechanical properties of LiF crystals.

It was shown that elementary radiation process when combined with various external fields differs essentially from the corresponding process under the same irradiation conditions but without these fields. Particularly, the additional (apart from crystallographic) anisotropy of mechanical and optical properties, caused by formation of oriented anisotropic radiation defects (bi-vacancies, F_2^+- centers etc.) and preferred slip planes, is induced.

The idea to combine various external fields was also used when studying post-radiation phenomena in crystals irradiated in reactor (annealing with uniaxial compression, hard UV irradiation with mechanical load etc.).

Results of magneto-plastic effect studies in crystals under combined action of weak magnetic field and X-raying, which demonstrate the role of spin-dependent transitions in the formation of post-radiation properties of crystals, are presented.

[1] E-mail: vkvachadze@yahoo.com.
[2] E-mail: maxsvet@yahoo.com.

INTRODUCTION

Advance of material science for nuclear and cosmic engineering, as well as various force majeure situations, requires development and production of new radiation resistant materials. For successful employment under extreme conditions (radiation fields of nuclear reactors of the new generation including) they should have better physical-mechanical characteristics as compared with materials currently in use. Problems of primary importance for development of engineering are as well: a) the problem of radiation sensitivity of materials that are used in computer engineering and dosimetry as storage units, and b) the problem of radiation involvement into technological processes used for the production of materials with specified physical properties.

The goal of the proposed review is to analyze the results, accumulated for more than 25 years in the E. Andronikashvili Institute of Physics in the sphere of radiation physics of solids, from the abovementioned standpoint and to show topicality of the available experimental data for the current problems of radiation material science. These results were obtained for solids under the combined action of ionizing radiation and various loads (thermal, electrical, magnetic, mechanic fields etc.). Such investigations are extremely important both from the fundamental (detection and identification of new radiation induced phenomena) and applied viewpoints, because experimental conditions to a large degree approach the conditions under which the material is actually to operate in various units.

1. MECHANICAL PROPERTIES OF LIF CRYSTALS EXPOSED TO THE COMBINED ACTION OF IONIZING RADIATION AND MECHANICAL (OR ELECTRICAL) LOAD

As is known, both radiation and mechanical action lead to yield stress increase (i.e. to hardening) and to plasticity decrease (i.e. to embrittlement), which restricts application of the crystalline materials practically important for operation under irradiation. Naturally, it was expected that combined mechanical and radiation action will lead to total hardening increase and plasticity decrease of crystals. However, it was shown both experimentally and theoretically that the simultaneous action of radiation and mechanical (or electrical) load leads not to additive hardening increase and plasticity decrease of the crystal, but to its hardening increase without embrittlement [1-4]. It was a quite unexpected result for the scientific notions of the time.

1.1. Stress - strain Diagram for LiF Crystals under Study

Samples sheared along cleavage planes {100} were neutron irradiated in the mechanically stressed state – uniaxial compression along [100] with stress σ insignificantly exceeding yield point of unirradiated sample σ_0 (σ_0=5.0 MPa).

Diagram, shown in Figure 1 demonstrates that irradiation of a crystal in a free state (curve 2) usually results in strong hardening increase and complete plasticity drop. Nearly complete recovery of ultimate strain takes place (curves 3 and 4) along with crystal hardening after irradiation of the crystal in the stressed state. So, a hard and plastic material was produced instead of a hard and brittle one. However, yield stress increment in this case is less than in the case of samples irradiated in a free state.

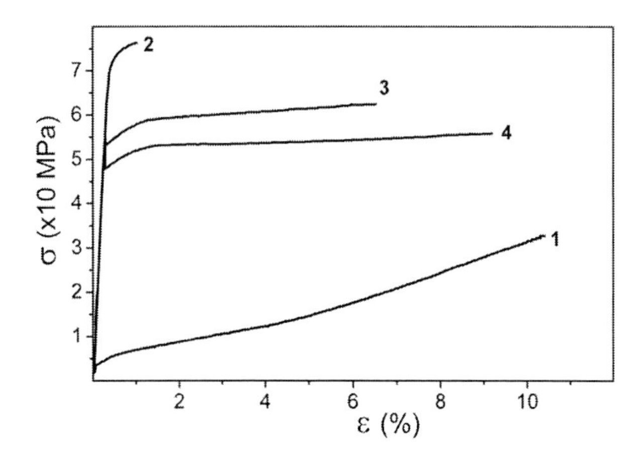

Figure 1. Stress – strain diagram for LiF crystal before (curve 1) and after irradiation (10^{15} n/cm^2) in a free state (curve 2) and under the load of 5.3 and 5.8 MPa (curves 3 and 4 respectively) [1].

In addition, slope of the curve in the plastic deformation area for samples irradiated under load (curves 3 and 4) does not change as it happens usually for unirradiated samples (curve 1), which gives evidence that dislocation motion in these samples takes place in one system of slip planes and_is the result of induction of the additional (apart from crystallographic) anisotropy of mechanical properties. The investigations of a deformed sample in the polarized light and by the chemical etching (Figure 2) confirm the assumption that slip bands belong mainly to one system of "easy" slip planes being in operation at post-radiation deformation of the crystal irradiated in a stressed state.

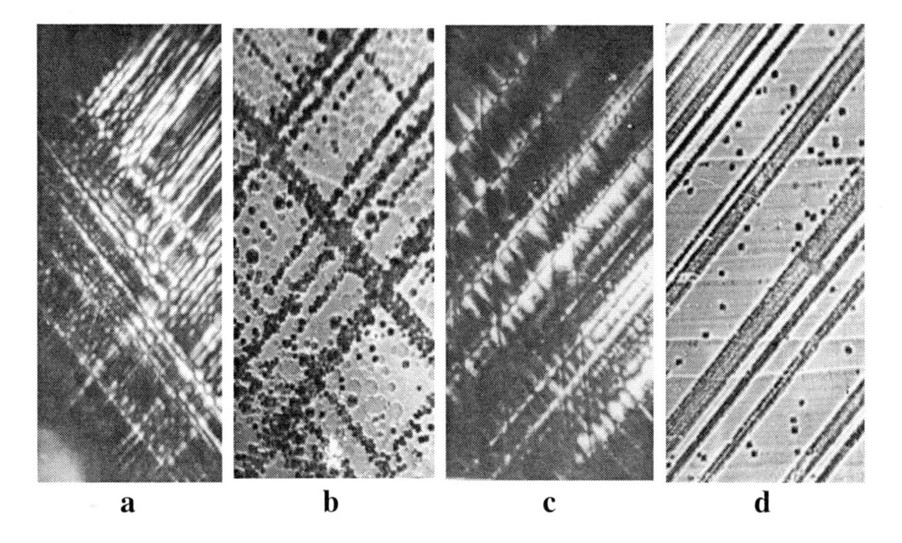

Figure 2. Polarized-light pictures (a and c) and etching patterns (b and d) of the deformed samples: a and b – samples irradiated in a free state (all equivalent systems of slip planes are operating); c and d – samples irradiated in a stressed state (one system of slip planes is operating) [2].

These data account for the increased plasticity of crystals stressed under irradiation (as compared with the crystals irradiated in a free state). At high stresses in crystals irradiated in a free state a significant (12-fold) increase of yield limit leads to the simultaneous activation of all four equivalent systems of slip planes. Interaction of dislocations different systems retards and locks motion of dislocations entailing the sample embrittlement [2].

1.2. Mechanism of Induced Anisotropy

To define the mechanism of induced anisotropy it is necessary to reveal the type of defects responsible for radiation hardening of LiF crystals. For this purpose temperature dependence of shear stress τ was studied in these crystals. A $\tau^{1/2}$ - $T^{1/2}$ plot of abovementioned dependency yields a straight line [3], which according to Fleisher theory [5] corresponds to interaction of dislocations with anisotropic tetragonal defects. Extrapolation of straight lines to the value $\tau = 0$ gives their intersection with the temperature axis at the same point, which testifies to the fact that shear stress τ for crystals irradiated both in a free and stressed states (under the fixed irradiation dose) is due to the interaction of dislocations with anisotropic defects of one type. Differing slope of the lines indicates a difference in the concentration of these defects, at this the more is the stress in the course of irradiation the less is this concentration.

Since interstitial atoms in ionic crystals anneal at low temperatures, among the anisotropic defects created in the crystal by neutron irradiation at normal temperatures bi-vacancies exceed in number, and the change of shear stress τ for irradiated LiF crystals is mainly caused by the interaction of dislocations with these bi-vacancies.

Bi-vacancies are as well the main defects making contribution to the hardening of crystals in the case of gamma irradiation. Facility, making it possible to load the sample along any direction of system <110> and to determine critical shear stress τ in a particular slip plane, was designed in house. Measurement results are given in Table.

Table. Shear stress τ in the system of slip planes {110} in LiF crystal

Irradiation type and intensity	Applied load	$\tau_{[10\bar{1}]}$	$\tau_{[01\bar{1}]}$
Reactor, 10^{15} n/cm^2,	$\sigma = 5.3$ MPa	43.0 MPa	24.0 MPa
γ - radiation, 7×10^5 rad,	$E_{[0\bar{1}0]} = 1{,}3 kV/cm$	16.7 MPa	10.6 MPa

From Table it is seen that for crystals irradiated by gamma-rays in electric field (field direction coincides with one of the possible orientations of bi-vacancies axis [010] - orientation 1, Figure 3), the dependence of τ on crystallographic direction is observed: plane of "difficult" slip is (101) plane, where the axis of defect is parallel to dislocation, and one of the "easy" slip planes is (011) plane. With the increase of electric field stress, the anisotropy of mechanical properties increases, as it is seen in Figure 4: shear stress τ for the plane of "easy" slip (011) decreases (curve 1), and for the plane of "difficult" slip (101) – increases (curve 2); both curves tend to saturation, which corresponds to saturation of the oriented defects concentration (τ_0 is the shear stress in the system of slip planes {110}of the crystal irradiated without field). The anisotropy caused by the application of electric field disappears within several hours after the field removal.

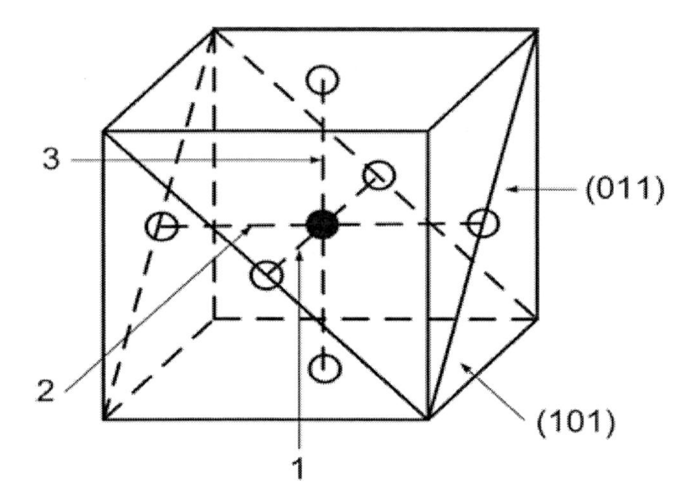

Figure 3. System of slip planes and orientation of bi-vacancies axis 1, 2 and 3 are shown in a structure of NaCl-type crystals [3].

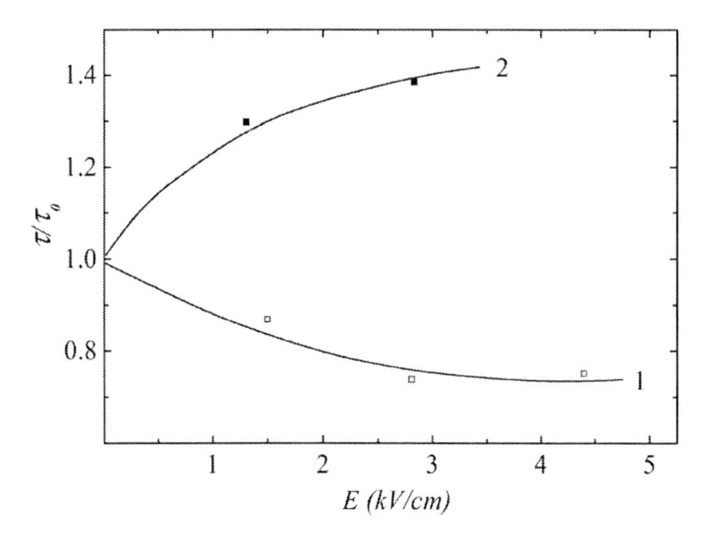

Figure 4. Dependence of shear stress τ on electric field intensity E, applied to LiF crystal in the direction of $[0\bar{1}0]$ in the process of γ-irradiation (10^6 rad): the plane of "easy" slip (011) corresponds to curve 1 and the plane of "difficult" slip (101) corresponds to curve 2 [4].

The similar anisotropy is observed for previously equivalent slip planes in the case of crystal irradiation in mechanically stressed state (see Table). If the irradiated crystal is under load corresponding to the first stage of strain hardening, one system of slip planes is active (in case under consideration the plane (101)). Near these planes anisotropic defects induced by irradiation take the orientation which is preferable because the energy corresponding to this orientation is lower as compared with other orientations. This means that defect population over all possible orientations is no more isotropic. Sample test after irradiation indicates that plane system, for which these oriented defects present the least resistance to the dislocation motion (see Table), proves to be active. From experimental (as well from calculated [6]) data it follows that in this case also the plane system, where defect axis is parallel to the dislocation line, is found to be "difficult". The induced anisotropy of properties caused by the above-described action is retained for a long time.

Thus, under conditions of combined action of radiation and mechanical stress (or electric field) anisotropic defects, formed in the crystal during irradiation, are oriented in the preferred direction, inducing additional anisotropy of mechanical properties of such crystals, which in its turn ensures plasticity conservation.

The revealed property should be inherent not only to ionic crystals, where it was detected and studied, but also to other crystalline solids and should be of importance for studies of materials having practical importance (metals and alloys, semiconductors, including glasses, ceramics, polymers etc.). Actually, similar phenomena were to the equal measure observed for metals, both in model and real (intra-reactor) tests [7, 8].

Plasticization of crystals with oriented anisotropic defects was employed to increase the optical resistance of NaCl crystals commonly used as the elements of optical system in CO_2 lasers. It proved to be that gamma-irradiation of crystals (10^6 rad) in a stressed state permits to increase optical resistance of the crystal under the action of laser beam 2 or 3-fold as compared with the reference sample [9]. The character of crystal destruction has also changed: under the action of laser beam the crystal irradiated in a free state cracks more (Figure 5a) than the gamma-plasticized crystal (Figure 5b). It is obvious that in plastic crystals, dislocations easy slip accounts partially for dissipation of energy delivered into the crystal by the laser beam in such mode preventing the brittle failure of crystal.

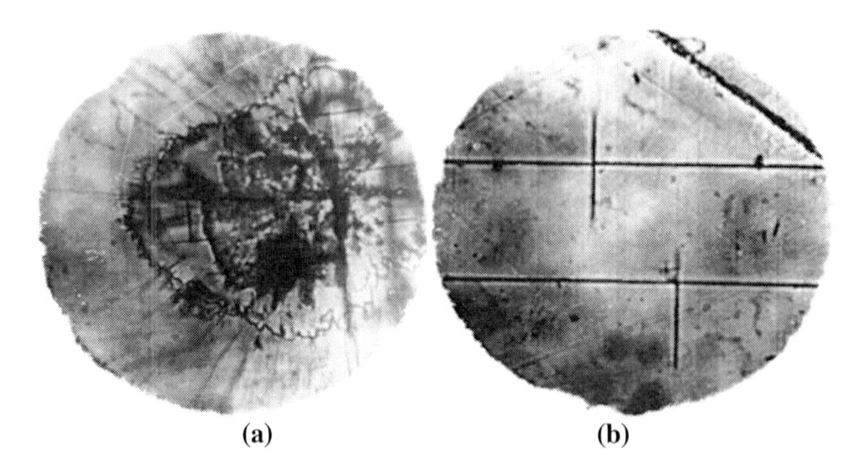

Figure 5. Photomicrography of action area of CO_2 laser beam on the surface of NaCl crystal γ-irradiated (10^6 rad) in a free (a) and mechanically stressed states (b).

1.3. Post-radiation Annealing of the Crystals under Mechanical Load

Additional anisotropy can be induced in a crystal not only during the process of its irradiation in a stressed state, but also during post-radiation annealing of irradiated crystals with mechanical load applied [10].

Figure 6 shows the etching pattern of irradiated LiF samples (10^{18} n/cm^2) annealed at 450^{0}C under the conditions of uniaxial compression along [001]. One can see the etching pits corresponding to dislocation loops. Their possible orientations along <100> and <110> are schematically shown in Figure 6. Redistribution of loops with differing orientation caused by the load is distinctly seen (c and d).

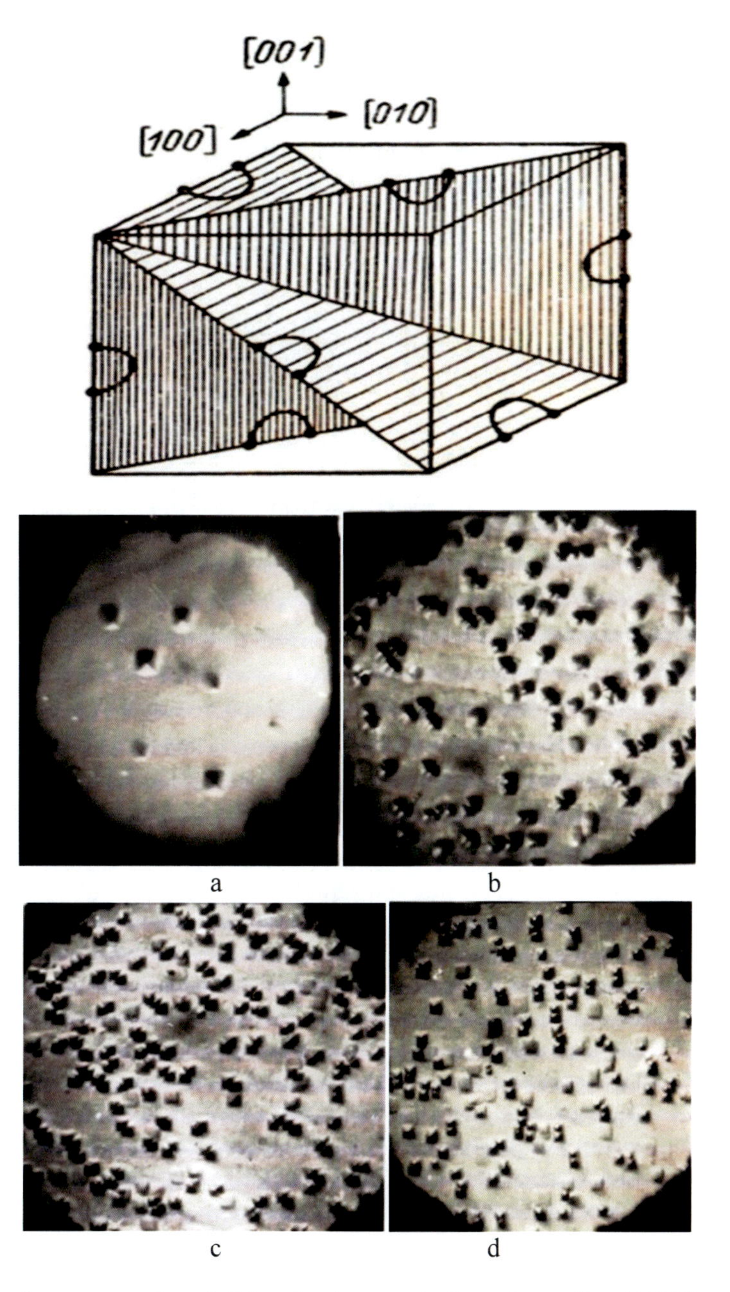

Figure 6. Schematic representation of dislocation loops and effect of annealing (450°C) on irradiated LiF crystal: a – before annealing; b – annealing in a free state (loops are oriented along <100> and <110>); c and d – annealing in a stressed state: c – surface perpendicular to load (loops are oriented only along <110>); d – surface parallel to load (loops are oriented only along <100>) [10].

Analysis of this dislocation structure and its evolution under combined action of temperature field and mechanical load, as well as performed calculations made it possible to conclude [10] that the dislocation loops formed at condensation of point defects are loops of the interstitial type.

In the medium oversaturated with vacancies there appear gas pore nuclei, which at high-temperature annealing can grow to the dimensions easily seen by the optical microscope. Indeed, the heating of irradiated samples above 600^0C revealed the micro-cavities which are pores (Figure 7a), oriented along <100> direction with linear dimensions in the range of 1-50μm and with the thickness of ≤ 0.4μm. The pores are retained in crystals up to the melting temperature. Combination of temperature field and mechanical load provides specific orientation to the pores in a manner similar to that of dislocation loops (Figure 7b) [11].

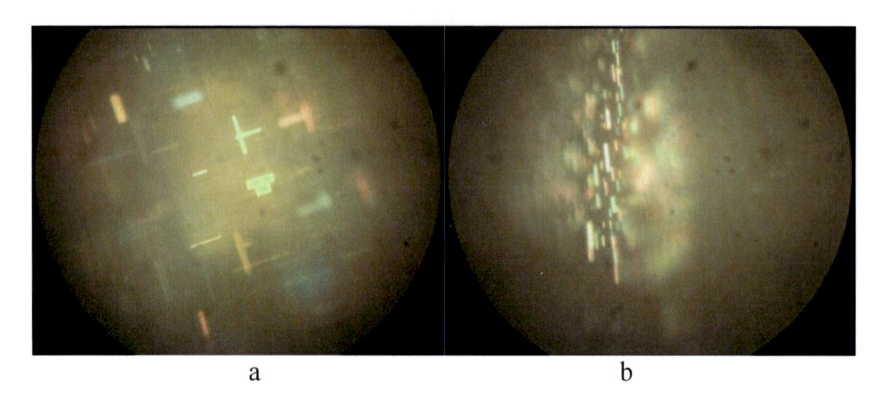

a b

Figure 7. Microcavities in irradiated LiF crystal after annealing in a free (a) and mechanically stressed (b) state [11].

2. Optical Properties of LiF Crystals Exposed to Combined Action of Ionizing Radiation and Mechanical Load (Creation of Stable F_2^+- Centers)

Colored alkali halide crystals (AHC) are extensively used for creation of tunable lasers on color centers (CC) [12, 13]. Such lasers (e.g. LiF with F_2^+-centers) jointly with lasers on different CC overlap wide range of generation frequencies – from 0.7 to 4 micron [3]. Unfortunately, low thermal stability of F_2^+- centers at room temperature (in LiF half-life ≤ 12 h) made active medium of AHC laser unusable, as it could be operated only at quite low temperatures (< 100 K [14]) thus making both laser construction and its operation more complicated.

Impurity ions of bivalent metal and/or hydroxyl decay products [15-20] are very important for the increase of F_2^+-centers thermal stability. Both impurity types ($F_2^+O^{2-}$ and $F_2^+V_c^-$) not only form stable electron traps and additional anionic vacancies but also promote emergence of various perturbing defects in the vicinity of F_2^+-centers. However, high impurity

[3] Lasers based on solutions of organic dyes, with tuning in near UV are extremely bulky.

concentrations as well as high irradiation doses deteriorate main generating characteristics and optical resistance of crystals [21].

Problem of F_2^+-centers stabilization acquires primary importance in connection with the abovementioned. Generation of stable at room temperature (RT) F_2^+-color centers in nominally pure LiF crystals (without dopants) was performed by combined action of ionizing irradiation and mechanical stress. Three groups of experiments were performed:

- gamma-irradiation of LiF single crystals in a stressed state ($\sigma \approx \sigma_0$, where σ is the applied mechanical stress, σ_0 is yield stress of virgin unirradiated sample) [22-24];
- action of hard (pulsed) UV-radiation and shock wave on LiF single crystals colored by irradiation [24, 25];
- deformation (up to $\sigma \approx \sigma_0$) of colored by irradiation LiF single crystals in the course of UV bleaching.

Thus, generation of stable at RT F_2^+-centers was studied not only during radiation coloration of crystals but also during "reverse" processes – bleaching of crystals colored by radiation and decay of radiation complexes into simpler formations.

2.1. F_2^+-centers in LiF Crystals Colored in a Stressed State

Optical absorption spectra of preliminarily annealed LiF samples registered immediately after gamma-irradiation (~0.5Mrad) at RT in a free (curve 1) and stressed states (curves 2 and 3) are given in Figure 8. In the latter cases only pronounced F_2^+ absorption bands (645 nm) are observed. Redundant, as compared with samples irradiated in a free state, accumulations of F-(250 nm) and F_2-centers (450nm) are also observed [22-24]. Evidently, in the stressed crystals there arise favorable conditions for separation of Frenkel defects (F and H), formed in the course of irradiation. On the one hand, this promotes generation of redundant F-centers and, on the other hand, increases formation probability of F-aggregate centers (F_2 and F_2^+).

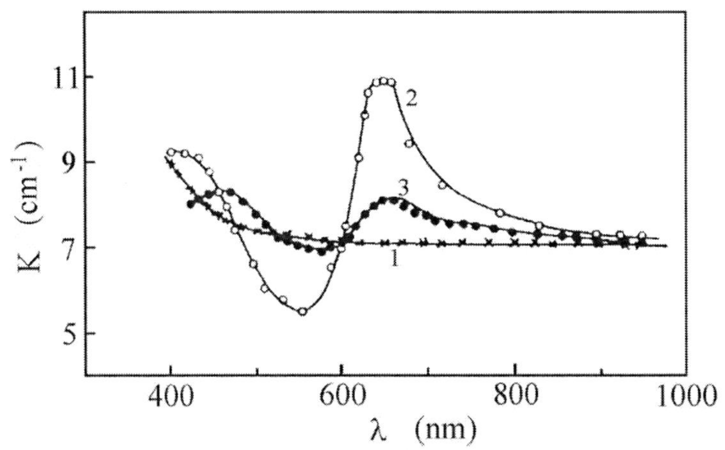

Figure 8. Optical absorption spectra of LiF crystals irradiated in free (curve 1) and stressed states at $\sigma = 2.0$ MPa (curve 2) and $\sigma = 2.5$ MPa (curve 3). Gamma-irradiation dose is ~ 0.5 Mrad [22-24].

Effect of deformation upon formation of F_2^+-centers decreases with applied irradiation dose and mechanical stress increase. Dependence of absorption coefficient at the maximum of F_2^+-band on the value of the applied mechanical stress [24] under constant irradiation dose is given in Figure 9. The maximal effect of stable F_2^+-centers accumulation is observed in the case when value of the applied mechanical stress does not exceed significantly yield stress of the crystal under study i.e. when new dislocations are not formed and an abrupt change of the existing order in the sample does not take place. Effect of accumulation of F_2^+-centers stable at RT strongly decreases with transition to the plastic region of deformation (see also curves 3, Figure 8). The similar phenomenon takes place under big irradiation doses when large aggregations of radiation defects are formed.

Thermal stability of the obtained F_2^+-centers proved to be very high and in optical absorption spectrum there was a small maximum of F_2^+-band even after 20 months' storage in dark at RT [24]. Dependence of detected effect on temperature and irradiation intensity as well as sample's microstructure was studied [22, 23].

Evidently, external mechanical load can exert considerable effect on the process of radiation-induced defect formation only in the case when the internal stress in the greater part of the crystal is substantially lower than the stress arising in the crystal due to the applied load. The elastic stress, generated in the crystal by dislocation, is inversely proportional to the distance from dislocation axis r,

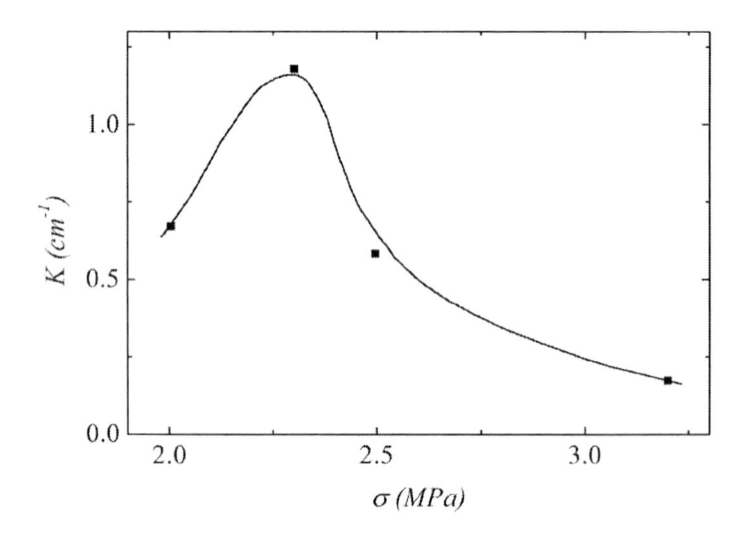

Figure 9. Dependence of optical absorption coefficient at the max of F_2^+ band on the applied mechanical stress value. Gamma-irradiation dose is ~ 0.5 Mrad [24].

$$\sigma(r) \approx G\frac{b}{r}, \tag{1}$$

where G is the shear modulus of crystal, and b is the value of dislocation Burgers' vector. If external mechanical load creates in the crystal stress σ', then from condition $\sigma' = \sigma(r_D)$ it is possible to determine radius of cylindrical area around dislocation r_D, where the internal stress is undoubtedly bigger than σ', and then determine relative part of crystal volume λ, where $\sigma(r) > \sigma'$,

$$\lambda = \pi r_D^2 \rho = \pi (G/\sigma')^2 b^2 \rho \quad (2)$$

Setting λ equal to one, it is possible to determine the critical density of dislocations,

$$\rho_k = 1/\pi r_D^2 = (\sigma'/G)^2 (1/\pi b^2) \quad (3)$$

which is the upper limit of ρ values, when it is still possible to observe effect of combined action of external stress and irradiation.

If the stress caused by external load is less than yield stress σ_0 (in our case $\sigma' < \sigma_0 \approx 2.0 \div 2.3$ MPa), then using $b = 4\times10^{-10}$m and $G = 4\times10^{10}$ N/m^2, it is possible to derive the estimation $\rho_k \leq 10^{10}$ m^2.

Indeed, the above-described effect of external load, applied to a crystal in the course of irradiation, can be distinctly traced in annealed crystals ($\rho < 10^{10}$ m^{-2}), irradiated with $\sigma' \approx \sigma_0$. In crystals irradiated with $\sigma' > \sigma_0$, when dislocations density is increased by more than an order of magnitude, this effect disappears almost completely regardless of whether crystal deformation was performed in the course of irradiation or before it. It should be added that separate (by turns) action of gamma-irradiation and mechanical stress upon LiF crystals did not entail formation of stable configurations of F_2^+-centers. Moreover, post-radiation deformation of preliminarily irradiated sample did not cause any change in the range of optical absorption spectra under study (200 – 1100nm).

Thus, the combined action of considerably small doses of gamma-irradiation (0.5 Mrad) and moderate mechanical stress ($\sigma \approx \sigma_0$) upon nominally pure LiF crystals results in considerable increase of F_2^+-centers lifetime at RT.

2.2. Hard (Pulsed) UV Irradiation and Shock Wave Effect on LiF Crystals

It is known that one of the ways to create F_2^+-centers is two-step ionization of F_2 centers ($F_2 + h\nu \rightarrow F_2^* + h\nu \rightarrow F_2^+ + e$). It is usually achieved by action of UV irradiation upon preliminarily radiation colored LiF crystals. However, only UV irradiation is not sufficient to obtain thermally stable F_2^+-centers. In [25, 27, 28] it was shown that creation of thermally stable F_2^+-centers is possible in the case of combined action of hard (pulsed) UV radiation and mechanical stress in a shock wave[4] mode. Optical absorption spectra of crystals exposed to such joint action (discharge voltage is 15 kV) is given in Figure 10.

[4] Treatment of LiF colored crystals by hard pulsed UV and a shock wave was performed on unit where creeping discharge on a surface of dielectric (modification of "plasma plateau") was used as a source of UV radiation and shock waves [27].

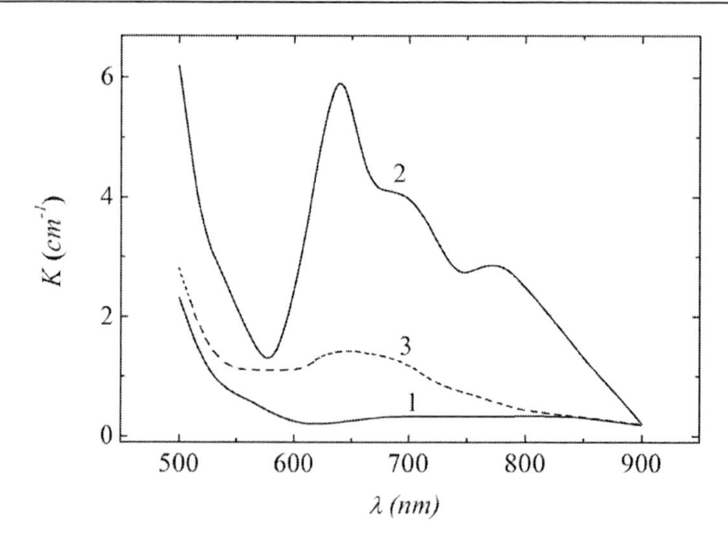

Figure 10. Optical absorption spectra of irradiated LiF crystals (2×10^{14} n/cm^2) before (curve 1) and after the action of pulsed UV and shock wave directly (curve 2) and through a quartz plate (curve 3) [24, 25].

Band with max at 645nm is the result of F_2^+-centers formation (curve 2). Maxima at 678 and 786 nm on the same curve correspond to $F_3^-(R_1,R_2)$ centers. Action of pulsed UV radiation only (quartz plate protects the sample from shock waves) abruptly decreases the efficiency of stable F_2^+-centers creation (curve 3).

The aforesaid combined action promotes not only creation of stable F_2^+ and F_3^- centers, but also more effective accumulation of other F-aggregate color centers, both charged F_3^+ (458 nm) and uncharged F_2 (450 nm), F_3 (380 nm) and F_4 (550 nm). Concentration of F-centers also increases as it was in the case of crystals irradiated under stress – absorption coefficient at the max of F-band before and after the action is respectively 191.4 and 216.2 cm^{-1}.

Without preliminary irradiation in reactor no change in optical absorption spectra of crystals (200 – 1000nm) is observed. Presence of certain amount of hydroxyl impurity ions (~ 10ppm and higher) in nominally pure crystals proved to be crucial for detection of the effect [25].

With discharge voltage increase up to 30kV (Figure 11) efficiency of stable F_2^+-centers formation decreases (discharge does not occur when voltage is below 15kV). Efficiency of their accumulation also decreases in the case of repeated combined action of hard UV radiation and shock wave on crystals under study, discharge voltage being constant [25].

Thermal stability of F_2^+- centers in radiation colored LiF crystals obtained by above-described method was found to be as high [25] as that in crystals first compressed and then gamma-irradiated (see 3.1). Luminescence spectrum registered after storage of the single crystal in dark at RT for 21 months (Figure 12b) exhibits emission band of F_2^+- centers with max at 950nm and spectral width of 65nm.

Curve of F_2^+- centers destruction (12a) indicates that we deal with F_2^+- centers of two types. The first type includes short-lived F_2^+- centers which decompose mainly in the course of one-and-half or two weeks after exposure of the crystal to UV with shock wave; the second type centers are stabilized configurations with F_2^+- centers, amount of which slowly decreases in the course of one year and more.

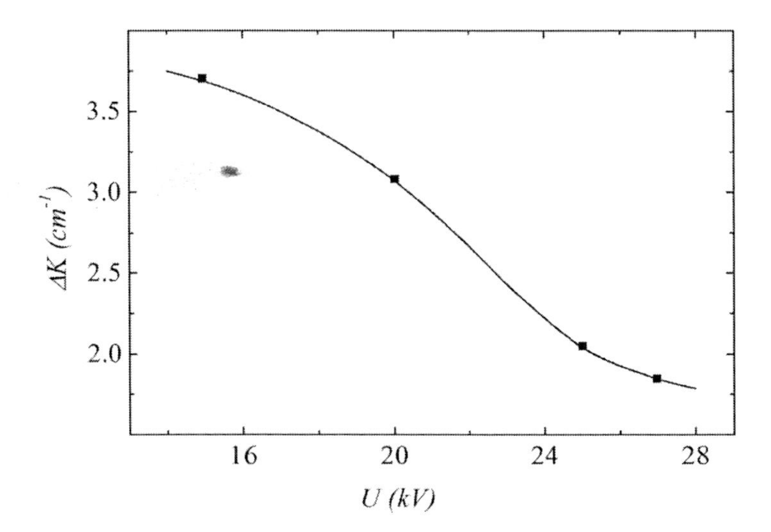

Figure 11. Dependence of **optical absorption coefficient changes** (ΔK) **at** F_2^+ band max on discharge voltage.

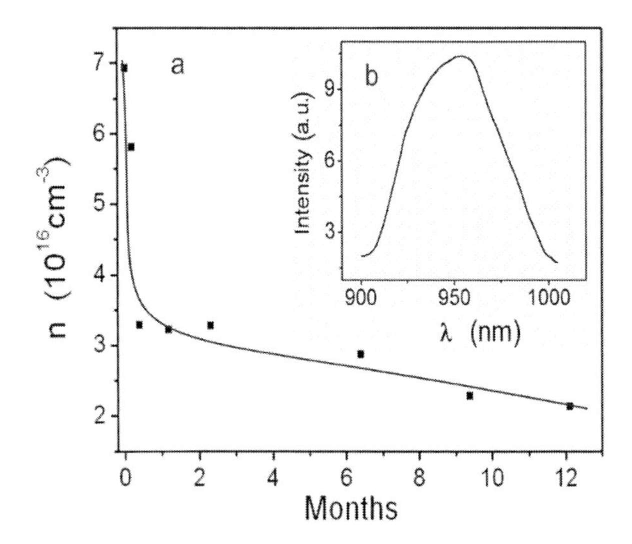

Figure 12. Dependence of F_2^+ center concentration on crystal storage time in the dark at RT after exposure to hard UV and shock wave (a) and luminescence spectrum after 21 months of the aforesaid action (b) [25].

Curve of F_2^+- centers destruction (12a) indicates that we deal with F_2^+- centers of two types. The first type includes short-lived F_2^+- centers which decompose mainly in the course of one-and-half or two weeks after exposure of the crystal to UV with shock wave; the second type centers are stabilized configurations with F_2^+- centers, amount of which slowly decreases in the course of one year and more.

2.3. Mechanical Load of Irradiated Crystals (up to $\sigma \approx \sigma_0$) in the Process UV Photo-bleaching

It was shown in 3.1 that irradiation of LiF crystals by neutrons or gamma-rays in a stressed state increases life-time of F_2^+ color centers usually unstable at room temperature. The applied mechanic all stress in the region of the yield stress of virgin sample was constant during the whole period of irradiation, whereas the yield stress as a rule increased due to dislocation pinning by radiation-induced defects. The similar experiments were carried out in post-irradiation regime as well, when preliminarily irradiated crystals were first loaded mechanically and then exposed to UV photo-bleaching.

In [26] formation and stabilization of abovementioned CC in LiF single crystals, irradiated in reactor (neutron fluence ~ 5×10^{15} n·cm^{-2}), were studied by means of combined action of UV photo-bleaching and continuously increasing mechanical stress (up to σ_0 of sample under irradiation).

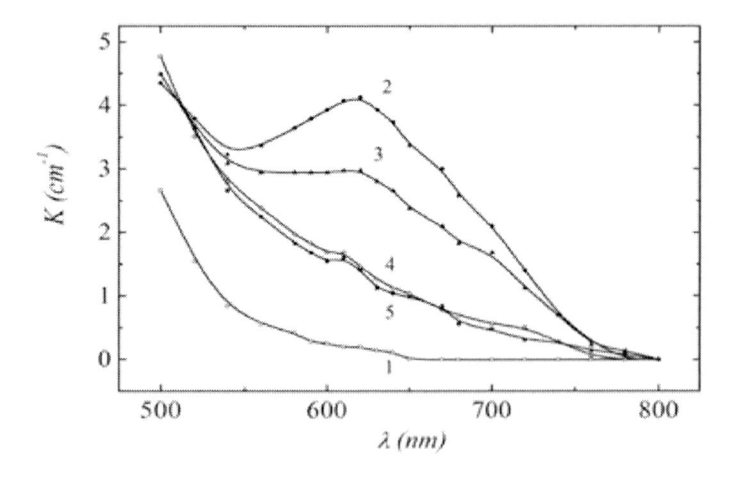

Figure 13. Optical absorption spectra of irradiated LiF crystals before (curve 1) and after combined action of UV and mechanical load (curve 2), as well as after storage in the dark at RT for 5, 72 and 100 hours (curve 3, 4 and 5 correspondingly) [26].

Figure 13 shows optical absorption spectra (500-800 nm) of LiF samples before (curve 1) and after photo-bleaching with continuous loading in the region of the yield stress (curve 2, wave length of UV light did not exceed 366nm), as well as after storage in the dark at RT (curves 3-5 registered 5, 72 и 100 hours after UV action). The absorption band (620 nm) characteristic of F_2^+ centers is observed, its intensity decreasing in the course of time.

The trend of F_2^+ band intensity decrease with time is presented in Figure 14. The experimental points, measured at the maximum of F_2^+ band (periodically registered on the sample kept in the dark), well fit curve, which is the combination of two exponents. One of them shows a rapid decrease of intensity (unstable F_2^+ centers), and the second exhibits a slow decrease (F_2^+ centers stabilized during the experiment [25]). Thus, what is again at issue is the presence of two types of F_2^+ centers in the crystal under study. The experimental points for the reference sample irradiated by UV-light in a free state without mechanical stress fit one exponent indicating the presence of one-type F_2^+ centers (unstable F_2^+ centers) in the reference sample.

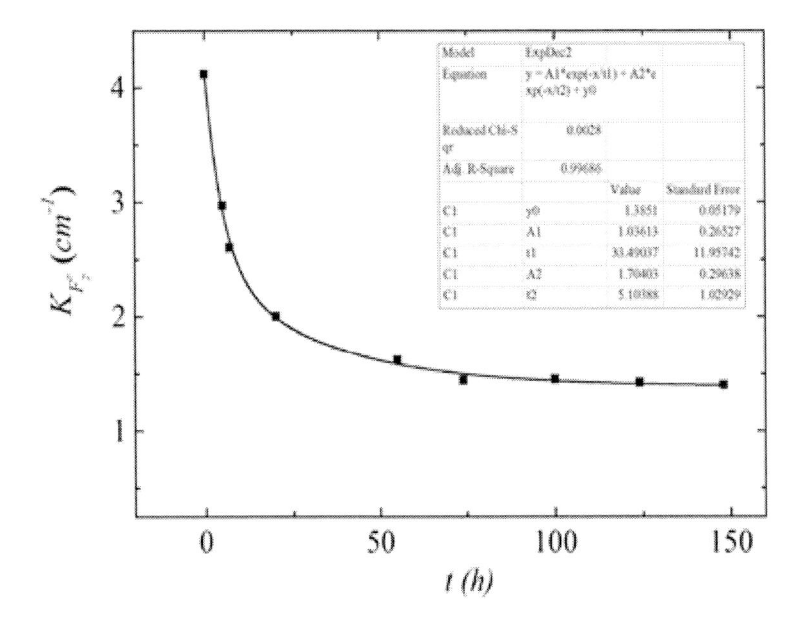

Figure 14. A number of experimental dots characterizing dependence of absorption coefficient (F_2^+ band maximum at 620 nm) on the storage time of irradiated LiF crystal, subjected to combined action of UV and mechanical load. The fitted curve is the combination of two exponents [26].

The luminescence curve of the crystal displaying the region of the induced emission of F_2^+ CC after the combined action of UV light and deformation is given in Figure 15 (curve 2). Curve 1 in the same figure corresponds to the sample irradiated in reactor prior to aforesaid action.

After the deconvolution of the luminescence curve two peaks are well pronounced: the first one is in the range of 860-920 nm, where the main luminescence peak is observed (curve 3); the second peak appears at approximately 920 nm (curve 4).

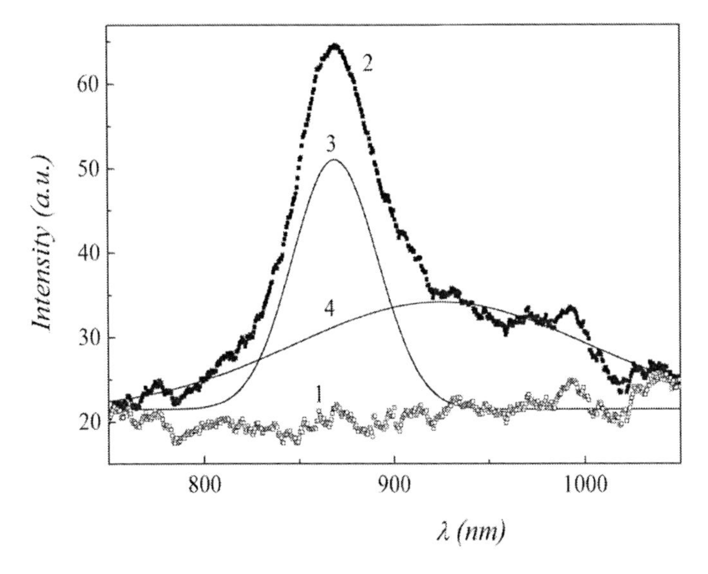

Figure 15. Luminescence spectra of irradiated LiF crystals before (curve 1) and after combined action of UV light and mechanical load (curve 2); curves 3 and 4 are the result of curve 2 deconvolution [26].

In our opinion, the second luminescence region is due to the stabilized F_2^+ CC. The wavelength of this emission maximum (about 920 nm) indicates that these CC are $F_2^+O^{2-}$ centers (O^{2-}-$Va^+ + F \rightarrow F_2^+O^{2-}$ and/or $F_2^+ + O^- + e^- \rightarrow F_2^+O^{2-}$ [20]) which have the maximum of the absorption band at ~645 nm [25]. A small plateau on curve 2 (Figure 13) in the vicinity of 650nm gives good cause to suppose (*for the sample*) existence of such band which does not manifest itself because is overlapped by more intensive band with max at 620 nm. This assumption makes results of optical absorption and luminescence even more adequate. So, after aforesaid combined action F_2^+ centers of two types are observed in the crystal: unstable F_2^+ centers (their absorption and luminescence maxima are 620 and 870 nm respectively) which are destroyed mainly during first 24 hours (Figure 14) and stabilized configurations with F_2^+ centers ($F_2^+O^{2-}$), which are retained in the crystal for 140 hours and more (their absorption and luminescence maxima are 650 and 920-950 nm respectively).

The change of the deformation rate during the combined action of UV light and mechanical load has little or no effect on the concentration of F_2^+ CC that are already formed, but it affects the degree of their stability. Figure 16 shows the dependence of K_{F2+} on the deformation rate, where K_{F2+} is the value of absorption coefficient of $F_2^+O^{2-}$ centers within 15 hours after the combined action. It is seen that with the increase of deformation rate, the fraction of stabilized complexes with F_2^+ centers ($F_2^+O^{2-}$) in the total amount of F_2^+ centers increases.

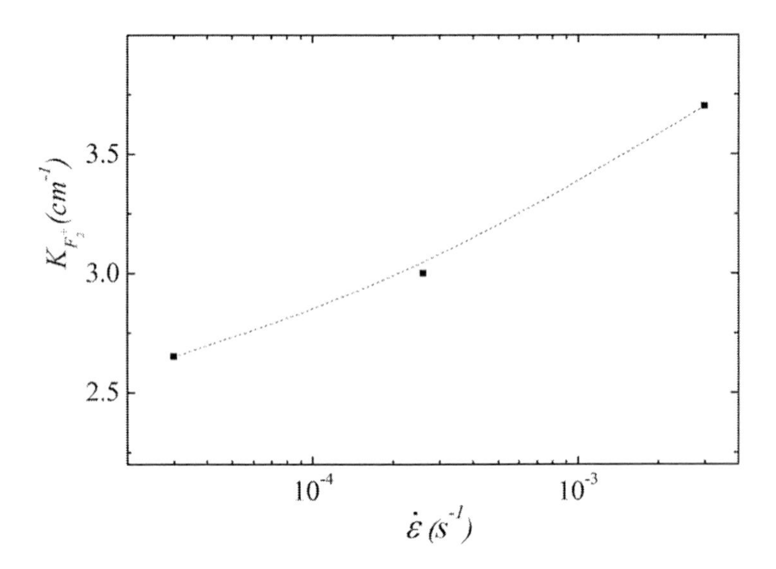

Figure 16. Dependence of absorption coefficient (F_2^+ band maximum at 620 nm) on the deformation rate of the irradiated LiF crystal, subjected to combined action of UV light and increasing mechanical load [26].

The first results on induced emission generation (in pulsed mode) in radiation colored crystals exposed to the combined action of UV and mechanical load [26] showed that generation efficiency of stable F_2^+ centers is higher than that of the unstable centers. This can be caused by anisotropic defects (including $F_2^+O^{2-}$ centers) of one preferable orientation that are created in crystals under study by aforesaid actions.

2.4. On Creation Mechanism of Color F_2^+-centers Stable at RT in Nominally Pure LiF Crystals

Impurity ions of bi-valent metals or hydroxyl decay products are very important for stabilization of F_2^+-centers. According to frequency location and other characteristics of F_2^+-band observed in [29] one can conclude that stabilizer of F_2^+-centers is the product of hydroxyl decay oxygen, which has not been introduced into the crystal intentionally and is present in small amounts. There are substantial number of papers which show that oxygen incorporated into LiF lattice as anion substituting ion O^{2-}, forms dipoles $O^{2-} - V_a^+$ and more complicated aggregates of centers and affects properties of radiation-induced defects. Evidently, complex of F_2^+-center with oxygen can have configurations that differ from each other both by energy of binding [29] and energy of reorientation activation.

It is not excluded that the most stable configuration does not occur when straying complex components meet occasionally (as barrier exists), and that activation energy of complex transition from one configuration to other is so high that thermoactivated transitions are of low probability. This means that stable configurations of these complexes cannot arise on the second (diffusion) stage of radiation-induced defect formation. They can be created only at the first dynamical stage e.g. during radiationless decay of electron excitation in the close vicinity of oxygen-vacancy complex with subsequent trapping of electron.

Naturally, frequency of such events will increase when number of oxygen-vacancy complexes goes up. The latter can be attained by special doping of the crystal by hydroxyl which is necessary when producing crystals with stable F_2^+-centers by irradiation in a free state [29]. Whereas in crystals under study amount of impurities able to stabilize F_2^+-centers is considerably lower and density of stable F_2^+-centers created under irradiation in a stressed state approximately the same, one can conclude [23] that anisotropic field of mechanical stresses with clearly preferable direction (which is created in the crystal by uniaxial load) considerably increases probability to form dynamically stable configuration of complex: F_2^+-center - oxygen impurity.

It is not excluded that in a asymmetric stresses field the barrier of complex formation, being formation with its own anisotropic elastic field, decreases in some directions to such an extent that its surmount in the radiation-excited lattice is considerably facilitated. This will increase the efficiency of F_2^+-center junction with oxygen ion and can lead to formation of considerable amount of $F_2^+O^{2-}$ centers even in the case when concentration of oxygen ions is so small that it is impossible to detect the formation of these centers as a result of only ionizing radiation without external stimulation. Such assumption is quite realistic if one takes into account calculations in [8,30]. According to these results mechanical stress applied to the sample induces changes of the intracrystalline potential relief that in turn affects the value of dynamic crowdion runaway i.e. amount of substituent defects and departure distances between vacancies and interstitial atoms.

Effect of asymmetric stresses arising due to uniaxial load can be essential only in the case when they are higher than the internal *stress*, which can be caused by structural defects existence, dislocations particularly. In the vicinity of dislocations internal stresses are much higher than those caused by external loads and in some spaces these stresses can stimulate stabilization of F_2^+-centers. However, in annealed crystals due to the low values of dislocations density (~10^{10} m^{-2}) share of spaces with such stresses is too small and average level of internal stresses in the crystal on the whole is very low.

If crystal is loaded above yield stress, plastic deformation is initiated and dislocations density rapidly increases, which leads to increase of the internal stresses level. But these stresses, as they represent superposition of fields generated by dislocations, are distributed randomly both in space and in directions of axis and the Burgers vectors, are nearly isotropic and cannot promote formation of stable complexes as do not have necessary orientation. On the other hand, due to their high level internal stresses can reduce efficiency of external field effect and decrease accumulation of stable F_2^+-centers.

To demonstrate, dependence curve of absorption coefficient at F_2^+-band max on applied load (see Figure 9) goes through maximum when applied load equals the yield stress. In what follows, efficiency of stabilized F_2^+-centers accumulation considerably decreases with applied stress increase. As it was mentioned, decrease of stable F_2^+-centers concentration was also observed at discharge voltage increase i.e. at increase of combined action efficiency (hard UV radiation and shock wave). Evidently, in this case also considerable increase of internal stresses level occurs.

Dependence of F_2^+centers accumulation on multiplicity of UV and shock wave action can also be explained in the same way. In other words, shock wave (as well as pulsed electric field [24,25]) applied to colored LiF crystals in the course of UV irradiation plays the same role as external load applied to the crystal under irradiation, creating necessary anisotropy in the crystal. It is exactly because of this anisotropy that correlation between the results of these different groups of combined actions experiments is observed.

If uniaxial compression at $\sigma \approx \sigma_0$ were to affect the diffusion stage of radiation-induced defect formation, then the external stress could have lead to the orientation ordering of radiation-induced defects and the above-described effects would have occurred also in the case of crystal post-radiation loading i.e. radiation would have created anisotropic defects and then uniaxial load would have ordered them according to the orientation. However, in our experiments such variant was not observed.

The result presented in Figure 16 is completely consistent with the concept. The thing is that the induced anisotropy is most strongly revealed at the high deformation rates, when the deformation takes place in one system of slip planes [31]. Besides, it follows from the concept that the accumulation efficiency of the stable F_2^+ centers should increase, which is confirmed experimentally.

3. Magnetoplastic Effect in LiF Crystals X-Rayed in a Weak Magnetic Field

In connection with the above-stated studies of ionizing radiation and magnetic field combination [32, 33] seemed if not necessary then interesting, particularly with reference to magnetoplastic effect (MPE) [34-36], which consists in the change of plastic properties of diamagnetic crystals in a weak magnetic field (MF).

Physical model of MPE in nonmagnetic crystals is based on the concept of spin-dependent electron transitions in the external magnetic field: the field enables some spin transition which radically changes configuration in the system dislocation-paramagnetic center (stopper), which entails decrease (or increase) of local barriers for dislocation motion.

Necessary condition for MPE is the availability of nonequilibrium defects, X-raying being one of the ways to create them. Both creation of new defects and change of state of defects already existing by this time in the virgin crystal [37] are meant. Consequently, it could be presumed that combination of X-raying and MF, on the one hand, would increase MPE and, on the other hand, would enable detection of MF effect on the amount and/or distribution of magneto-sensitive and other defects.

After preliminary tests of virgin crystals (MPE strongly depends on the impurity content [37], down to the effect suppression), samples were exposed to combined action of MF and X-rays. After removal of external action measurements of yield stress were performed, the results are shown in Figure 17.

With irradiation times up to t=15 min samples irradiated in MF exhibit lower yield stress than samples irradiated without MF; maximal difference is about 25% and decreases to zero with dose increase - in ~ 10^3s curves σ_0 (t) merge. Thus, MF effect on the radiation hardening of crystal is observed only at small doses of X-raying. This result (as well as data of other authors [37]) testifies to the fact that new magnetosensitive defects are 'born' at the earliest stages of X-raying.

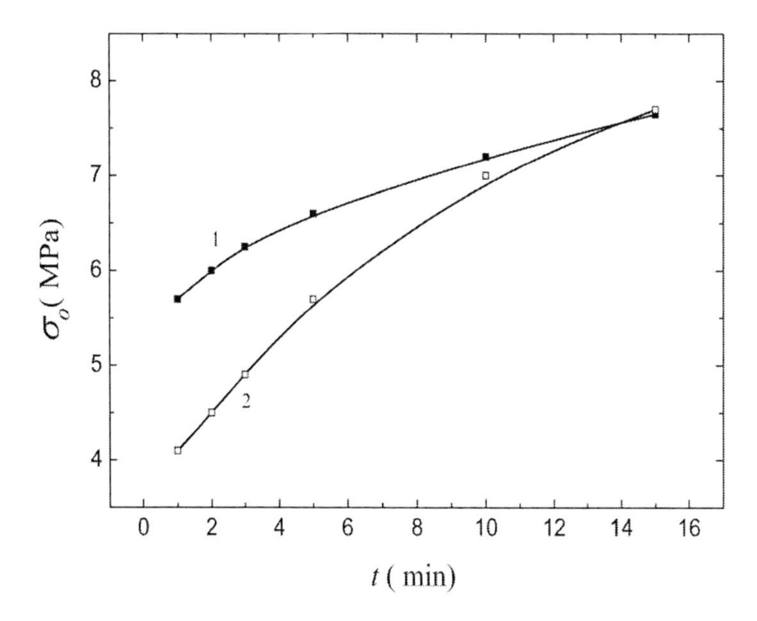

Figure 17. Dependence of yield stress σ_0 on the time of X-raying: irradiation without MF (curve 1) and in MF (curve 2).

The fact, that at the beginning of irradiation the process leading to loss of strength prevails (Figure 17) and then with the radiation change of microstructure this effect decreases to zero, indicates existence of two types of potential stoppers for dislocations motion in crystal under irradiation. The first type comprises intrinsic radiation-induced defects and the second one contains impurity complexes that existed initially in the crystal and were changed by magnetic field in the process of irradiation; as it was shown in [37] irradiation makes impurity more magnetosensitive. The latter can lead to reduction of local barriers for dislocations movement and therefore to yield stress decrease. With dose increase concentration and hardening role of intrinsic radiation-induces defects increases, whereas role of impurity complexes diminishes. In the long run competition of these two processes reduces to zero total effect of MF on radiation hardening of LiF crystals in ~ 10^3s time.

Under given conditions no display of induced anisotropy was observed.

CONCLUSION

Summing up the aforesaid we come to a following conclusion: combined action of radiation and mechanical load (or electric field) upon LiF crystals results in the additional (apart from crystallographic) anisotropy of mechanical and optical properties, caused by orientation of anisotropic radiation-induced defects by applied external field and interaction of these defects with dislocations. This induced anisotropy facilitates considerably:

- increase of yield stress of the crystal without its embrittlement (nonadditive effect with respect to results of separate action of radiation and load);
- increase of life-time of laser F_2^+-centers in nominally pure LiF crystals (usually unstable at room temperature) without their predoping by cations or anions, which deteriorates main generating characteristics of active element and decreases optical resistance of crystal.

The suggested combination of external actions on the contrary increases optical resistance of crystals.

Quite different nature of ionic crystals plastification is displayed in case of irradiation in weak magnetic fields. Here spin-dependent processes manifest themselves, opening up new possibilities in plasticity physics of nonmagnetic crystals.

REFERENCES

[1] Andronikashvili, E. L.; Galustashvili, M. V.; Dokhner, R. D. In *Defects in Insulating Crystals (Proceedings of the International Conference*); Tuchkevich, V. M.; Svarts, K. K.; Ed.; Publisher: Riga, Latvia; Berlin Heidelberg New York, 1981, pp 439-458.

[2] Paperno, I. M.; Galustashvili, M. V. *FTT* 1976, vol 18, 259-261.

[3] Paperno, I. M.; Galustashvili, M. V. *FTT* 1976, vol 18, 1941-1943.

[4] Andronikashvili, E. L.; Paperno, I. M.; Galustashvili, M. V. *FTT* 1978, vol 20, 1423-1425.

[5] Fleisher, R. *J. Appl. Phys.* 1962, vol 33, 3504-3508.

[6] Andronikashvili, E. L.; Dokhner, R. D. *FTT* 1981, vol 23, 298-302.

[7] Gindin, I. A.; Naskidashvili, I. A.;Lapiashvili, E. S.; Nekljudov, I. M. *Problemi prochnosti* 1973, 8, 49-52. (in Russian, Strength problems).

[8] Japaridze, S. K.; Trushin, Ju.V. *JTF* 1985, vol 55, 1244-1246.

[9] Andronikashvili, E. L.; Paperno, I. M.; Galustashvili, M. V.; Barkhudarov, E. M.; Taktakishvili, M. I. *FTT* 1979, vol 21, 2739-2743.

[10] Andronikashvili, E. L.; Galustashvili, M. V.; Driaev, D. G.; Saralidze, Z. K. *FTT* 1987, vol 29, 130-135.

[11] Abramishvili, M. G.; Galustashvili, M. V.; Dekanozishvili, G. G.; Driaev, D. G.; Kalabegishvili, T. L.; Kvatchadze, V. G. In *New Developments in Material Science;* Chikoidze, E.; Tchelidze, T.; Ed.; Publisher: New York, USA, pp 169-176.

[12] Mollenauer, L. F.; Olson, D. N. *J. Appl. Phys.* 1975, vol 46, 3109-3118.

[13] Basiev, T. T.; Mirov, S. B. *Room Temperature Tunable Color Center Lasers*, Harwood Acad. Publisher: Chur. Switzerland, 1994, vol. 16, pp. 1-160.

[14] Mollenauer, L. F. In *Tunable Lasers*; Mollenauer, L. F.; White, J. C.; Ed.; Berlin, FRG, 1987, Vol. 59, 225-277.

[15] Lobanov, B. D.; Maksimova, N. E.; Khulugurov, V. M.; Parfianovich, I.A. *FTT* 1980, vol 22(1), 283-285.

[16] Mollenauer, L. F. In *Defects in Insulating Crystals (Proceedings of the International Conference);* Tuchkevich, V. M.; Svarts, K. K.; Ed.; Publisher: Riga, Latvia; Berlin Heidelberg New York, 1981; pp 524-541.

[17] Lobanov, B. D.; Maksimova, N. T.; Isianova, E. D.; Lomasov, V. N.; Prokhorov, A. M.; Tcirulnik, A. M. *Optics and Spectroscopy* 1987, vol 63(4), 816-822.

[18] Sierra, R. A.; Collins, C. R. *Appl. Opt.* 1982, vol 21(24), 4400-4405.

[19] Courrol, L. C.; Ranieri, M. *Phys. Rev.* 1990, vol B42(7), 4741-4747.

[20] Dergachev, Yu.; Mirov, S. B. *Optics Communications* 1998, vol 147, 107-111.

[21] Basiev, T. T.; Voronko, Ju. E.; Mirov, S. B.; Osiko, V. V.; Prokhorov, A. I. Proceedings of the Academy of Sciences of the USSR 1982, *Series Physics, vol.* 46(8), 1600-1608.

[22] Kvachadze, B. G.; Abramishvili, M. A.; Altukhov, V. I. *J. Low Temp. Phys.* 1985, vol 58(1/2), 143-152.

[23] Abramishvili, M. G.; Kvachadze, V. G.; Saralidze, Z. K. *Solid State Physics* 1987, vol 29(1) 39-43.

[24] Abramishvili, M. A.; Akhvlediani, Z. G.; Barkhudarov, E. M.; Kalabegishvili, T. L.; Kvachadze, V.A .; Taktakishvili, M.I . In *Proc. Conference Laser Physics*; Vartanian, E.C.; Ed.; Ashtarak, Armenia, 1996, 95-102.

[25] Abramishvili, M. G.; Akhvlediani, Z. G.; Kalabegishvili, T. L.; Kvachadze, V. G.; Saralidze, Z. K. *Solid State Physics* 1998, vol 40(11), 2044-2050.

[26] Abramishvili, M.; Abramishvili, G.; Galustashvili, M.; Kalabegishvili, T.; Kvatchadze, V. J. Luminescence 2012, *LUMIN-D-11-01069.*

[27] Abramishvili, M. A.; Akhvlediani, Z. G.; Barkhudarov, E. M.; Kalabegishvili, T. L.; Kvachadze, V. A.; Taktakishvili, M. I. *FTT* 1995, vol 37(8), 2526-2528.

[28] Abramishvili, M. G.; Akhvlediani, Z. G.; Kalabegishvili, T. L.; Kvachadze, V. G.; Saralidze, Z. K. *Solid State Physics* 1998, vol 40(11), 2044-2050.

[29] Parfianovich, I. A.; Khulugurov, V. M.; Ivanov, N. A.; Titov; Ju. M.; Chepurnoi; Ju. M.; Varnavskii; O. P.; Shevchenko; V. P.; Leontovich, A. M. *Izvestia Acad. Sci. USSR, Ph.Ser.* 1981, vol 45(2), 309-314.

[30] Japaridze, S. K.; Trushin, Ju.V. In *Computer simulation of crystal defect structure;* LFTI AN USSR; Publisher: Leningrad, USSR, 1987, 202-204.

[31] Galustashvili, M. V.; Driaev, D. G.; Saralidze, Z. K. *Phys. Solid State* 1985, vol 27(8), 2320-2324.

[32] Galustashvili, M. V.; Abramishvili, M. G.; Driaev, D. G.; Kvatchadze, V. G. *Physics of the Solid State* 2011, vol 53 (7), 1412-1414.

[33] Galustashvili, M.; Abramishvili, M.; Driaev, D.; Kvatchadze, V.; Tsakadze, S. In *Nano Studies;* Chkhartishvili, L.; Ed.;. Publisher: Tbilisi, Georgia, 2011; Vol. 3, 199-204.

[34] Alshits, V. I.; Darinskaya, E. V.; Koldaeva, M. V.; Petrzhik, E. A. In *Dislocations in Solids;* Hirth, J. P.; Ed.; Publisher: Amsterdam, Netherlads, 2008; Vol. 14, 333-437.

[35] Golovin, Yu.I. *FTT* 2004, vol 46, 769-803.

[36] Morgunov, R. B. *UFN* 2004, vol 174, 131-152.

[37] Alshits, V. I.; Darinskaya, E. V.; Perekalina, T. M.; Urusovskaya, A. A. *FTT* 1987, vol 29, 467-471.

In: Ionizing Radiation
Editors: Eduard Belotserkovsky and Ziven Ostaltsov ISBN: 978-1-62257-343-1

Chapter 5

MTDNA DELETIONS AND POSSIBLE EMPLOYING OF THEM AS MARKERS IN RADIOLOGICAL INVESTIGATIONS

V. N. Antipova*[*], *L. V. Malakhova*, *A. M. Usacheva and M. G. Lomaeva
Institute of Theoretical and Experimental Biophysics,
Russian Academy of Sciences, Pushchino, Moscow Region, Russia

ABSTRACT

Most studies of radiation-induced changes in genetic material of cells are based on an analysis of nuclear DNA. Lately there has been growing interest of researchers in the state of mitochondria and their genome in tissues after genotoxic influences on the organism, in as much as the stability of mitochondrial DNA (mtDNA), carrying genes of the system of energy biogenesis, is extraordinarily significant for the processes of cell restoration. MtDNA in the functioning and organization of structure has a number of peculiarities distinct from nuclear DNA (nDNA). It is replicated independently of the cell cycle and replication of nDNA [1]. In mitochondria there is limited functioning of various systems of DNA repair [2]. Apart of that, mtDNA does not contain noncoding sequences, in the structure of its genes there are no introns, and it is transcribed as a unified polycistronic block [1, 3]. In mtDNA there arise significantly more damages than in a commensurate fragment of nDNA, both as a result of influence of reactive oxygen species (ROS) generated in mitochondria themselves and upon the action on the cells of physical and chemical genotoxic agents. All this favors formation in mtDNA of disturbances/mutations at a high frequency (see review [4]). Therefore, some researchers use mtDNA as a marker of radiation damage by analyzing changes in the number of copies, presence of deletions and mutations in mtDNA in tissue cells, and amount of mitochondrial genome fragments in the plasma / serum [5-9].

Keywords: Mitochondrial DNA, mitochondrial DNA deletion, ionizing radiation

---

[*] Institute of Theoretical and Experimental Biophysics, Russian Academy of Sciences, Pushchino, Moscow region, 142290 Russia; E-mail: valerya@rambler.ru.

1. Introduction

It is known that exposure to physical and chemical environmental factors in small and moderate doses on human and animal organisms can cause alterations in the genome of somatic tissues, which may result in mutational and carcinogenic consequences. Currently, extensive use of nuclear technology and sources of ionizing radiation in various fields of industry leads to radiation exposure of a substantial proportion of the people population. Therefore, there is a need for a search for reliable genetic markers to evaluate the acceptable limits of radiation exposure on the tissue cells, as well as to assess the manifestation of long-term genetic effects of radiation in humans.

Most studies of radiation-induced changes in the genetic material of cells are based on an analysis of changes in the parameters of the nuclear DNA. Some researchers suggest using alterations in mtDNA as a marker [5, 6, 7, 8, 9]. This choice is based on at least two factors, namely high variability of mtDNA in nature and presence of multiple copies of the mitochondrial genome in cells [10, 11, and 12].

The first factor is apparently a consequence of low efficiency repair systems in mitochondria, which also makes mtDNA more susceptible to various chemical agents and ionizing radiation. The second factor allows the mitochondrial genome to persist in despite of the permanents effects of ROS [4]. Each cell containing a number of mtDNA molecules in the mitochondria can function even at very high levels of constant mtDNA damage due to repopulation of the mitochondrial genome on the basis of the remaining copies of "wild type" [13]. Therefore, some researchers use mtDNA as a marker by analyzing changes in the number of copies, presence of deletions and mutations in mtDNA in tissue cells, and amount of mitochondrial genome fragments in the plasma / serum.

This work focuses on such structural rearrangements of mitochondrial DNA as large deletions and considers potential of using such changes as a marker in radiological investigations.

2. The Structure of MtDNA Damage and Mutagenesis

The circular mitochondrial genome contains 37 genes; they include 13 of the structural genes encoding subunits of respiratory chain complexes, 22 tRNA genes and 2 rRNA genes, required for intra-mitochondrial synthesis of protein. In addition, there is a small 1.1. kb non-coding region called displacement or D-loop that contains essential sequences for the initiation of replication and transcription. Strands of mitochondrial DNA are asymmetric in composition pairs GC. Heavy (H - heavy) strand contains 28 genes. Light strand (L - light), contains a sequence of 9 genes [14, 1]. MtDNA replication proceeds bidirectionally on both strands. Fragment forms triple-stranded D-loop in the control region during the process of mtDNA replication [14].

To date, found that the mitochondrial DNA in the mitochondria of somatic cells are organized in nucleoids associated with proteins, involved in regulation of transcription and translation of mtDNA, and with histone-like proteins.[15]. As well as nuclear histones, histone-like proteins can shield mitochondrial molecules from the expose to various agents. [16].

Replication of damaged mtDNA, in contrast to that of nDNA, is not blocked by the system of cell cycle control; this promotes formation of mtDNA point mutations and deletions with a high frequency. This is a primary cause of heteroplasmy [2,17].

Heteroplasmy is defined as the presence of mutant and normal types of mitochondrial DNA copies within a mitochondrion, cell or an individual [18].

Although pathogenic mtDNA point mutations are almost invariably homoplasmic, the majority of pathogenic mutations in human mtDNA are heteroplasmic [19].In the process of cell division mitochondria are randomly distributed between daughter cells, which can result in both high levels of mtDNA mutations in some cells and a low level of it in other cells. However, the transfer of mtDNA to daughter cells can occur with the selection of mutant molecules. [20]. In addition, the degree of heteroplasmy within the cells of one tissue can vary throughout the whole life of the individual [21]. Damaged mitochondria with high levels of DNA mutations are capable of proliferation in postmitotic tissue cells by a yet unknown mechanism and thus increase the contribution of mutant mtDNA. In opposite, in rapidly dividing cells, mutant mtDNA is usually found in relatively lower ratio [22, 23; see review 24]. The severity of energy metabolism disorders in the cell is directly proportional to percentage of mutant mtDNA copies in its mitochondria. Thus, heteroplasmy and replicative selection often leads to the fact that the cells or individuals with the same nuclear genome (eg, identical twins) may have different mitochondrial genotypes and, as a result, differ phenotypically [25].

For the manifestation of mitochondrial dysfunction it is sufficiently that the number of copies of mtDNA mutations has exceeded a certain level - this phenomenon is called the threshold effect [19].

The sensitivity threshold depends not only on the mtDNA mutation type, its location, and the proportion of mutant mtDNA copies in the cells, but also on individual consumption of dimensional cells and tissues [27, 28; see review 24]. Perhaps that is why organs with high energy requirements are at higher risk rates of developing pathologies, which are referred to as "mitochondrial" disease [25]. But changes in the proportions of mutant mtDNA in dividing cells, such as skeletal muscle cells and neurons, cannot be explained by mitotic segregation. It is assumed that the cause of these changes may be permanent replication of mtDNA [29]. The appearance of mitochondrial dysfunction in tissues contributes to the development of various pathologies, including neurodegeneration, neuromuscular and cardiovascular disease, renal dysfunction, and endocrine systems. Abnormalities in the structure of mtDNA are characteristic of aging, Alzheimer's, Parkinson's, Huntington's, diabetes and other diseases. [30].

3. Mitochondrial DNA Deletion

One of the important mutational events (rearrangements), arising in the mitochondrial genome, exists as formation of large (extended up to several thousand base pairs) deletions, leading to a loss of some genes and, consequently, to a certain deficiency of the electron transport chain proteins. [31]. Deleted molecules of mtDNA, which retained the sequences of replication origin, are capable of repopulating faster than normal molecules of mtDNA [32]. This may lead to exceeded threshold of rearranged molecules level and, as a consequence, to impairment of the process of oxidative phosphorylation [33, 34]. With the formation of mtDNA deletions they associate some «genetic» diseases of human [30, 35 and 36]. In studies carried out on cell culture, formation of large mtDNA deletions after exposure to ionizing radiation was shown [7, 37 and 38].

3.1. Characteristics of MtDNA Deletions

The largest number of single deletions ranging in size from 2 to 8.5 kb located in the region of 11 kb in length (Large arc) between two points of replication origin (O_H and O_L) [39]. This site is more susceptible to deletions [40], perhaps because it remains single-stranded for a long time in the process of replication. However, the deletions were found in all regions of mtDNA [41].

Greater number of deletions (~ 85%) is flanked by short direct repeats. [42]. Therefore, the presence or absence of repetitive sequences is suggested as a characteristic of break points of deletions and, therefore, as a genetic criterion for the classification of deletions.

Classification systems Mita et al. [43] and Degoul et al. [44] are used to characterize the types of deletions by arranging deletions in two classes. To classify the types of deletions that are found in mice, Chung et al. [45] modified these systems by adding a third class. Class I includes deletions whose breakpoints directly flanked by perfect or degenerate direct repeats, which are localized in the normal mtDNA along the edges of the potential deletions. In these deleted genomes, one of the direct repeats is retained and starts immediately after the breakpoint, forming a deletion, while the other direct repeat remains in the deleted (remote) site.

Deletions of class I are specific: almost in all cases, the same repeat was found at the 5'-end, just before the "left" breakpoint of deletion, whereas the other repeat was found in a normal, nondeleted mtDNA at the extreme 3 '- end of deletion, just before the "right" break point of deletions. Direct repeats range in size from 5 to 13 bp. In the deleted genomes of class II direct repeat sequences were found near the breakpoints, but not at the exact nucleotide breakpoint positions.

Small direct repeats (perfect or imperfect) is often observed in the general area of breakpoints, but in no case was a repeat pair found exactly at the 5'-and 3'-ends of the deletions, which is the rule for the class I deletions. Class III includes deletions, which have no any evident repeat elements found near breakpoints.

Recently types of mtDNA deletions been studied, which are identified in the neurons of the aging human brain, in patients with Parkinson's disease and individuals with disabilities in the form of multiple deletions [46]. The results showed that the characteristics of mtDNA deletions are similar, with slight differences in size, presence or absence of flanking repeat sequences or in the length of the repeats. These observations allow us to expect that the mtDNA deletions are probably generated by a similar mechanism in different clinical situations [47].

3.2. Mitochondrial DNA Deletions and Mitochondrial Diseases

As indicated in the work [47], mtDNA deletions have an important role in human pathology in three different clinical scenarios. First, the single mtDNA deletions are a common cause of sporadic (or occasionally inherited from the mother), mitochondrial diseases, and in these cases, the same (identical) deletion (one and the same) can be found in all cells within the affected tissue) [48]. It is believed that there is clonal expansion of the deleted form of mtDNA during early development. Clonally expanded "single" deletions are often present in various forms, which can be interchanged by homologous recombination [49, 50].

Second, another group of individuals with mitochondrial diseases have multiple mtDNA deletions in various affected tissues, particularly in the muscles and central nervous system [30, 47]. Even so individual cells or segments of muscle fibers can contain only one or a small number of clonal expansions of rearranged forms of mtDNA [51]. It is believed that the primary genetic defect in these individuals affects nuclear genes encoding proteins required for mitochondrial nucleotide metabolism or for maintenance of mtDNA [52].

Third, there are numerous reports of mtDNA deletions in ageing postmitotic tissues of patients with neurodegenerative diseases. The extent of mtDNA deletions in these cases is often much lower than that seen in individuals with mitochondrial disease, but in substantia nigra neurons, for example, approximately 50% of all mtDNA molecules contain an mtDNA deletions [18,33,34]. The first evidence of damage in mtDNA in aging process has been obtained as a result of the detection of large deletions in mtDNA and the progressive increase in their number in various tissues of adult humans [24].

The copy number ratio of mtDNA with mutations versus wild-type mtDNA is a crucial factor in determining whether there will be consequences for biochemistry of the cells. As mtDNA mutations are functionally recessive, deletion and mutant load exceeding 60% of total mtDNA are usually needed to cause an overt biochemical defect [47, 53].

3.3. The Mechanisms of the Formation Mitochondrial DNA Deletions

At present, the molecular mechanism responsible for the occurrence of multiple mtDNA deletions is not known. The most discussed topics in the literature are the formation of deletions through replication errors [54] and by formation of DNA double strand breaks during repair of damaged mtDNA [47, 55]. The assumption of the important role of double strand breaks reparation in the formation of mtDNA deletions was made in the work performed on the cells culture using ionizing radiation [56, 7]. Double strand breaks may occur because of direct damage of mtDNA by ionizing radiation and as a consequence of increased production of ROS [57].

Currently, most researchers consider that replication is the likely mechanism behind deletion formation. At present, there are two models explaining the mechanism of replication of mtDNA: a traditional strand-asynchronous displacement model [58] and the model that assumes the presence of leading and lagging circuit (coupled leading-lagging strand replication) [59] (see [47]). It is likely that both types of replication take place in accordance with the functional state of cells. Thus, the synchronous type of mtDNA replication is characteristic of cells in the stationary phase, while asynchronous induction - for cells that need to rapidly increase the number of mitochondria [60].

Within the strand-asynchronous model of mtDNA replication it was suggested two hypothetical mechanisms for the formation of deletions. These models are «illegitimate elongation» [61] and «slip-replication [62]. Published studies suggested other possible mechanisms for the formation of deletions. Rocher et al. [63], in a model of «pyrimidine content», suggested the formation of the triple helix of DNA between guanine-rich direct repeats in the newly synthesized heavy thread and base strand paired to homologous pyrimidine-rich direct repeat in the light strand. To explain the mechanism of deletions, which breakpoints are not flanked by repetitive elements, but they are within or close to palindromic sequences, a model of «palindromic sequence» is suggested. Palindromic sequences, pretty common in the mitochondrial genome, can be involved into the mechanism of deletions through the formation of hairpin structures [64].

Based on the fact that the majority of deletions is flanked by not long direct repeats, but small, of 2-4 nucleotides (eg, in the rat genome they flanked up to 55% of all deletions [65]) , Chung et al. [45] proposed «replication jumping» model as an explanation of mechanism of deletion formation. This model assumes that the delay in polymerization reaction (polymerase "stutters"), occurring in those parts of the matrix filaments, which are modified by oxidative damage (such as single-strand breaks), leads to replication errors.

Some researchers suggested that mtDNA deletions can be formed in the process of repair of double strand breaks [66]. According to Krishnan et al. [47], the formation of deletions is initiated by single-strand fragments of mtDNA. The 3 '→ 5' exonuclease activity displayed by the mitochondrial DNA polymerase gamma [52] in the repair of double strand breaks, leads to the formation of single-stranded regions of mtDNA. These liberated single-strand fragments (with 5'- and 3'- repeat sequences) may be incorrectly annealed to micro-homologous sequences [67], such as repetitive sequences (including homopolymer regions), other single-strand fragments of mtDNA, or within the region D-loop, leading to subsequent repair, ligation of double strands and the degradation of the remaining unprotected single strand. This can lead to the formation of an intact mitochondrial genome having a copy with both the 5'-and 3' repeats and deleted part of the genome. The process of generation of mtDNA deletions, involving double-strand breaks, has also been shown [68]. A survey of individuals with disabilities in the form of multiple deletions also confirms the involvement of the repair process in the formation of deletions. It is shown that in individuals with mutations in the gene encoding helicase Twinkle, or the gene encoding the mitochondrial DNA polimerase gamma, the formation of double-strand breaks, arising as a possible consequence of frequent delay in replication, is initial step in the formation of mtDNA deletions in the muscles cells [69]. It is important that the replication delay also occurs under normal conditions, and this could lead to double-strand breaks that require reparation. The presence of spontaneous linear full length mtDNA, in various tissue samples from patients suffering from autosomal dominant progressive external ophthalmoplegia, supports the hypothesis of involvement of double-strand breaks in the process of generations of mtDNA deletions [70].

In addition to the described briefly above basic models, the contribution of the secondary structure of mtDNA in the process of deletion formation is discussed in literature [71], as well as involving in this process of a number of DNA repair enzymes [72], and a combination of the all mechanisms.

3.4. Detection of MtDNA Deletions

Total DNA or mtDNA can be analyzed for possible mtDNA rearrangements by Southern blotting, long-range PCR and real-time PCR [30].

The method of blot hybridization (Southern blot) remains one of the basic and reliable methods to identify a complete set of the deleted molecules and determine their number. An important limitation of this approach is its lack of sensitivity and high quality requirements to the target DNA (the preparation of DNA should be well purified, intact enough and sufficient in quantity). But even following such requirements, issue remains, that mtDNA isolated from cells with a mutant phenotype, may have a too small population of rearranged molecules for effective analysis by this method [73].

To detect large deletions of mtDNA method of long extension PCR is rather convenient, because it allows amplifying a full-size mtDNA, and any of its sites, which have binding sites for the primers. Full-size fragment and a few rebuilt (the deleted) mtDNA molecules, ranging in size from 10 to 2 kb, which have sites of "landing" - a place for the hybridization of primers [74], are synthesized in the reaction. Thus, the PCR products contain a full-size ("wild type") and some copies of the deleted mtDNA, which are then separated by electrophoresis. The presence of deletions is defining by identification of additional products of smaller size in the gel. The number and size of deletions detected by PCR, depend on the choice of primers (only those deletions are identified, breakpoints of which are located near the site of hybridization of the primers) and the length of the amplified product of the wild type.

To detect and quantify of certain / specific mtDNA deletions, the method Real-Time PCR is used in practice [75, 76].

4. The Deletions in the Mitochondrial DNA Is As a Marker Detecting Effects of Ionizing Radiation

Some authors have noted the accuracy of the method of using of mtDNA deletions and point mutations as markers of radiation exposure in cells [6, 7], and of other genotoxic factors of chemical and physical nature. It was showed [77, 78] that single and multiple deletions can be used as markers of photoaging of skin after damage caused by ultraviolet A (UVA; 320-400 nm), since the frequency of deletions increases with higher doses of UV irradiation.

The Armed Forces Radiobiology Research Institute of United States tried to use the effect of elevated induction of mtDNA rearrangements in human peripheral blood lymphocytes in biological dosimetry [6, 79].

Studies carried out on cell cultures [7, 37] showed that the deleted mtDNA copies appear within 72 hours after irradiation with X-rays and their formation is associated with replication of mtDNA [38]. Studies [8, 80] revealed deletions and point mutations in mtDNA in human cells after 12-24 hours after gamma-irradiation.

Changes in mtDNA are found in human peripheral blood cells exposed to radiation or chemotherapy. Investigation of mtDNA in blood cells of patients with a diagnosis of breast cancer (21 women) in the course of radiotherapy and chemotherapy showed contrasting changes in both number of copies of mtDNA relative nDNA, and in presence of large deletions of mtDNA in blood cells in response to the genotoxic effects. Some copies of mtDNA of blood cells in patients with breast cancer contain large deletions, the frequency of which increased after chemo- radiotherapy sessions [81]. The detection of emergence of large deletions (size 4977 bp) of mtDNA in human peripheral blood lymphocytes along with the searching for chromosomal abnormalities, are used in biological dosimetry [6]. The mtDNA and 'common' deletion content in irradiated human peripheral lymphocytes may be considered as predictive factors of radiation toxicity after total body irradiation treatment [82].

Our studies have shown that exposure of mice to radiation doses of 2 and 5 Gy increases the content of copies of mtDNA with large deletions in the tissues of the brain and spleen. Number of copies of mtDNA with deletions in the cells of tissues in the post-radiation period depends on the dose and time after irradiation [83]. The level of common deletions (size 4977 bp) content of mtDNA in directly irradiated cells increased with dose and was higher in radiosensitive cells [84].

These results suggest that ionizing radiation effectively induce large mtDNA deletions and such changes could be useful as a marker in radiological investigations.

However, in individuals exposed to radiation during nuclear weapons tests in the Semipalatinsk region, the frequency of deletions of 4977 bp was not increased compared to that of the control group [85]. In a study performed on human hepatoblastoma cell culture, irradiated with a dose of 5 Gy, a deletion of mtDNA 4977 bp in size had been determined up to 10 days of post-radiation time [38]. In a work [84] it has been demonstrated that in a bystander response positive cell line (F11-hTERT) common deletions level was increased until 35 days after ionizing radiation then reduced back to control level by day 49. Our experiments have shown a decline of levels of mtDNA copies with deletions in the tissues of irradiated animals within the post-radiation period (by the 28th day after irradiation). This may be associated with cell death in somatic tissues and the elimination of degraded fragments from tissue to the blood [83].

It would seem that the radiation-induced mutations / deletions in mtDNA somatic mammalian cells *in vivo* will be maintained over a long period of post and give their clonal propagation [86] within the cells of tissues of irradiated animals. However, in our experiments, we observed a decrease in the number of copies of mtDNA with deletions after 8 days of post- irradiation time. This paper [87] have shown that deletion 5914 bp in the cells of the ear tissue of 4 months animal could not be detected after irradiation in a dose of 2 Gy and 5 Gy. Probably copies of mtDNA with such damage have been eliminated from the tissue. Our data are consistent with the results obtained in the work [85], where it was shown that in rat tissues irradiated with single doses of 0.5 and 1 Gy, the number of deletions after 6 months did not significantly differ from controls. The observed decrease in the number of copies of mtDNA with deletions in the tissues of irradiated mice in the post-radiation period, is possibly due to the elimination of damaged mtDNA molecules from the tissues, as a result of organelles mitofagy [88, 89] or due to containing mtDNA cell death [38].

Production of ROS is increased in cells containing copies of mtDNA with deletions. This in combination with low levels of ATP can stimulate the mechanisms of cell death [90, 91]. It is possible that such a change as large mtDNA deletions can decrease the activity of the respiratory chain and lead to mitochondrial dysfunction. This scenario is consistent with the work of [92], in which it was shown that skeletal muscles of patients with the syndrome of

Kearns Sayre (possessing copies of mtDNA with deletions in 4977) contained a very few copies of mtDNA.

There is currently no information available about the keeping time of induced mtDNA mutations in the cells after exposure to the damaging agents. Perhaps, much of mutant mtDNA copies in peripheral blood cells (for a certain period of time after the damaging action of chemical agents or radiation) are subject to elimination in the process of renewing the blood. Obviously, removal of damaged mtDNA molecules from the mitochondria plays an important role in preventing the accumulation of mutant molecules, and preserves the integrity of the mitochondrial genome. The required number of copies of mtDNA in the mitochondria, after the removal of damaged copies, is provided by replication of remaining intact mtDNA in these organelles, or not significantly damaged molecules. This compensatory induction of mtDNA replication has been demonstrated in cells after oxidant treatment [93] and after irradiation of animals with ionizing radiation [94]. As soon as cell death and autophagy of mitochondria are cellular **routine** turnovers, extracellular mtDNA fragments or intact molecules can be found in human blood plasma and other biological fluids. They are of considerable interest for use as markers of various pathologies [95]. It is possible that damaged mtDNA or fragments of the genome translocates from mitochondria into the cytoplasm, or large fragments of mtDNA move into the cytoplasm after the self-destruction of the mitochondria by autophagy (mitofagy). The work gives evidence of the possible selective mitofagy of mitochondria carrying mutant copies of mtDNA [35, 88, 89]. It is possible that the increase in extracellular levels of mtDNA with deletions in the blood plasma can be used in the analysis of radiation damage [96].

Conclusion

The literature data and the the results of our laboratories show that the analysis of mtDNA deletions in general is a reliable and simple quantitative technique to study radiation effects. However, as soon as the mtDNA deletions are eliminated with time from irradiated cells or tissues in post-radiation period, more accurate approach is required to use the mtDNA deletions as a marker for the cell damage assessment in longer periods after radiation exposure. There is uncertain issue about how long mutant copies mtDNA can be retained in different tissues cells of animals and humans after exposure to radiation. To make it clear, there is a need to analyze rearrangements (deletions and mutations) in mtDNA in more distant period of time after irradiation. The question about the use of mtDNA deletions as markers of radiation injury remains open to improvements.

There is not enough information in literature about formation of mtDNA rearrangements in tissues of animals exposed to ionizing radiation. Also data for the analysis of deleted mtDNA, which were obtained from the personnel, exposed to radiation accidents, cancer patients after radiotherapy, or the population of the radiation-contaminated areas, are contradictory. Therefore, it seems urgent to continue the study of gamma-radiation induced mutations and deletions of mtDNA at the whole-organism level.

References

[1] Shadel G.S., Clayton D. A. (1997) *Annu. Rev. Biochem.,* 66, 409-435.
[2] Gaziev A.I, Podlutskii A. Ya. (2003) *Tsitologiya,* 45 (4), 403.
[3] Taylor R.W., Turnbull D.M. (2005) *Nat. Rev. Genet.*, 6, 389-402.
[4] Gaziev A.I., Shaikhaev G.O. *Genetika* (2008) 44, 437.
[5] Wardell T.M., Ferguson E., Chinnery P.F. et al. (2003) *Mutat. Res.,* 525, 19-27.
[6] Prasanna P.G. et al. (2002) *Mil. Med.,* 167(2 Suppl), 10-12.
[7] Prithivirajsingh S. et al. (2004) *FEBS Lett.,* 571, 227-232.
[8] Murphy J. E. et al. (2005) *Mutat. Res*, 585, 127-136.
[9] Abdullaev S.A. et al. (2009) *Molecular Biology,* 43, 6, 990-996.
[10] Beckman K.B., Ames B.N. (1999), *Mutat.Res*., 424, 51-58
[11] LeDoux S.P., Wilson G.L. (2001), *Prog.Nucleic.Acid.Res.Mol.Biol.,* 68, 273-284.
[12] Dianov G.L. et al. (2001), *Prog.Nucleic.Acid.Res.Mol.Biol*., 68, 285-297.
[13] Chomyn A. et al. (1992) *Proc. Nati. Acad. Sci. USA*, 89, 4221-4225.
[14] Anderson S. et al. (1981) *Nature,* 290, 457-465.
[15] Iborra F. et.al. (2004) *BMC. Biol.*, 2, 9.
[16] Kutsyi M. P. et al. (2005) *Mitochondrion,* 5, 35-44.
[17] Stuart J.A., Brown M.F. (2006) *Biochim. Biophys. Acta*, 1757, 79-89.
[18] Holt I.J. et al. (1988), *Nature,* 331(6158), 717-9.
[19] Schon E.A. et al. (1997) *J. Bioenerg. Biomemb.,* 29, 131-149.
[20] Chinnery P.F. et al. (2000) *TIG*., 16, 500-505.
[21] Clayton D. A. (2000) *Experimental Cell Research,* 255, 4-9.
[22] Shoffner J. M. (1996) *Lancet,* 348, 1283-1288.
[23] Wei Y.-H., Lee H.Ch. (2003) *Advances in Clinical Chemistry*, 37, 83-128
[24] Тодоров И.Н. (2007) Рос.хим.ж. LI, 1, 93-106.
[25] Wallace D. C. (1992) *Ann. Rev. Biochem.,* 61, 1175-1212.
[26] Hsieh R. H. et al. (2001) *J. Biomed. Sci*., 8, 328-335.
[27] Leonard J.V., Schapira A. H. (2000) *Lancet,* 355, 299-304.
[28] Pang C. Y. et al. (1999) *J. Formos. Med. Assoc.,* 98, 326-334.
[29] Chinnery P. F., Samuels D. C. (1999) *Am. J. Hum Genet*., 64(4), 1158-65.
[30] Taylor R.W., Turnbull D.M. (2005) *Nat. Rev*., 6, 389–402.
[31] Cooper J.M., Mann V.M and Schapira A.H. (1992) *J. Neurol. Sci*., 113, 91.
[32] Diaz F. et al (2002) *Nucl.Acids Res.,* 30, 4626.
[33] Kraytsberg Y. et al. (2006) *Nat. Genet*., 38, 518.
[34] Bender A. et al. (2006) *Nat. Genet*., 38, 515.
[35] Krishnan K.J. et al. (2007) *Nucl. Acids Res*., 35 (22), 7399.
[36] Khrapko K., Vij J. (2009) *Trends Genet.,* 25 (2), 91.
[37] Kubota N. et al. (1997) *Radiat. Res.,* 148, 395-398.
[38] Wang L. et al. (2007) *Int. J. Radiat. Biol*., 83 (7) 433-442.
[39] Kogelnik A. M. et.al. (1998) *Nucleic Acids Res*., 26, 112–115.
[40] Wei Y.H. (1992) *Mutat. Res.,* 275, 145–155.
[41] Lee C.M. et al. (1997) *Free Radic. Biol. Med*., 22, 1259–1269.
[42] Samuels D.C. et al. (2004) *Trends Genet.,* 20, 393–398.
[43] Mita S. et al. (1990) *Nucl. Acids Res*., 18(3), 561-567.

[44] Degoul F. et al. (1991) *Nucl. Acids Res.,* 19(3), 493- 496.
[45] Chung S.S. et al. (1996) Age, 19, 117-128.
[46] Reeve A.K. et al. (2008) *Am. J. Hum. Genet.*, 82, 228–235.
[47] Krishnan K.J. et al. (2008) *Nat.Genetics*, 40(3), 275-279
[48] Schaefer A.M. et al. (2008) *Ann. Neurol.*, 63, 35–39.
[49] Moslemi A.R. et al. (1996) *Ann. Neurol.*, 40, 707-713.
[50] Poulton I. et al. (1993) *Hum. Mol. Genet.,* 2, 23-30.
[51] Rotig A. et al. (1995) *Hum. Mol. Genet.*, 4, 1327-1330.
[52] Hudson G., Chinnery P.F. (2006) *Hum. Mol. Genet.,* 15 (Spec. No. 2), 244–252.
[53] Porteous W.K. et al. (1998) *Eur. J. Biochem.*, 257, 192-201
[54] Holt I.J. et al. (2000) *Cell*, 100, 515–524
[55] Fukui H., Moraes C.T. (2009) *Human Molecular Genetics*, 18 (6), 1028–1036.
[56] Ikushima T. et al. (2002) *International Congress Series*, 1236, 331-334
[57] Imlay J.A, Linn S. (1988) *Science*, 240, 640-642.
[58] Robberson D.L., Clayton D.A. (1972) *Proc. Natl. Acad. Sci. USA*, 69, 3810–3814.
[59] Yasukawa T. et al. (2006) *EMBO J.,* 25, 5358–5371.
[60] Mazunin I.O. (2010) *Russian Journal of Genetics*, 46 (9), 1100-1101.
[61] Buroker N.E. et al. (1990) *Genetics,* 124, 157–163.
[62] Shoffner J.M. et al. (1989) *Proc. Natl. Acad. Sci. USA*, 86, 7952–7956.
[63] Rocher C. et al. (2002) *Mol. Genet. Metab.*, 76, 123–132.
[64] Trinh T.Q., Sinden R.R. (1993) *Genetics*, 134, 409–22.
[65] Gadaleta M.N. et al. (1992) *Mutat. Res.*, 27 (3-6), 181-193.
[66] Lakshmipathy U., Campbell C. (1999) *Nucleic Acids Res.*, 27, 1198–1204.
[67] Haber J.E. (2000) *Trends Genet.*, 16, 259–264.
[68] Srivastava S., Moraes C.T. (2005) *Hum. Mol. Genet.*, 14, 893–902.
[69] Wanrooij S. et al. (2004) *Nuc.Acids Res.*, 32, 3053–3064.
[70] Van Goethem G. et al. (1997) *Eur.J Neurol.*, 4, 476–484.
[71] Glickman B.W., Ripley L.S. (1984) *Proc. Natl. Acad. Sci. USA*, 81, 512-516.
[72] Bae Y.S. et al. (1988) *Proc. Natl. Acad.Sci. USA,* 85, 2076-2080.
[73] Moraes C. et al. (2003) *J.Molecular Diagnostics*, 5 (4), 197-208.
[74] Reynier P., Malthiery Y. (1995) *Biocem. Biophys. Res. Commun*, 217, 59-67.
[75] Blakely W.F. et al. (2002) *Mil. Med.,* 167, 16–19.
[76] Pogozelski W. K. et al. (2003), *Mitochondrion,* 2, 415–427.
[77] Ray A.J. et al. (2000) *J. Invest. Dermatol.*, 115(4), 674-679.
[78] Koch H. et al. (2001) *J. Invest. Dermatol.,* 117, 892-897.
[79] Blakely W.F. et al. (2001) *Radiat. Prot. Dosim.*, 97, 17–23.
[80] Maguire P. et al. (2005) *Radiat. Res.*, 163, 384-390.
[81] Malakhova L.V. et al. (2006) *Вопросы онкологии*, 52 (4), 398-403.
[82] Wen Q. et al. (2011) *Radiation Oncology*, 6, 133.
[83] Antipova V.N. et al. (2011), *Biophysics*, 56 (3), 423–428.
[84] Schilling-Totha B. et al. (2011) *Mutat. Res.* 716, 33– 39.
[85] Hamada A. et al. (2003) *International Congress Series*, 1258, 169–176.
[86] Nekhaeva E. et al. (2002) *Proc. Natl Acad. Sci.USA*, 99, 5521–5526.
[87] Antipova V.N., Lomaeva M.G. (2011) *Russian Journal of Genetics*, 47(3), 376–377.
[88] Skulachev V.P. et al. (2004) *Mol. Cell. Biochem.*, 256-257(1-2), 341-58.
[89] Mijaljica D. et al. (2007) *Autophagy*, 3, 4 – 9.

[90] Jou M.J. et al. (2005) Ann. N.Y. Acad. Sci.,1042, 221–228
[91] Liu C.Y. et al. (2007) *Mitochondrion*, 7, 89-95.
[92] Barthelemy C. et al. (2001) *Ann. Neurol.*, 49, 607–617.
[93] Lee H. C. et al. (2000) *Biochem. J.*, 348, 425-432.
[94] Malakhova L.V. et al. (2005) *Cell. Molec. Biol. Lett.*, 10, 592-603.
[95] Tamkovich S.N. et al. (2008) *Molecular. Biology,* 42, 12-23.
[96] Vasilyeva I.N. (2001) *Ann. N. Y. Acad. Sci.*, 945, 221-228.

In: Ionizing Radiation
Editors: Eduard Belotserkovsky and Ziven Ostaltsov
ISBN: 978-1-62257-343-1

Chapter 6

ANTIOXIDANT PROPHYLAXIS OF RADIATION STRESS

Vitaly K. Koltover[1,*], Vladimir G. Korolev[2] and Yuri A. Kutlakhmedov[3]

[1]Institute of Problems of Chemical Physics, Russian Academy of Sciences, Chernogolovka, Moscow Region, Russia
[2]Petersburg Institute of Nuclear Physics, Gatchina, Leningrad Region, Russia
[3]Institute of Cell Biology and Genetic Engineering, National Academy of Sciences of Ukraine, Kyiv, Ukraine

ABSTRACT

Once atomic power engineering has become a part of our everyday lives, the special precautions should be taken to reduce harmful effects of radiation for specialists in atomic industry as well as for people in the contaminated territories after atomic accidents. To defend the people in case of chronic radiation, novel radiation protectors, which would be non-toxic and suited to long-time applications as nutrients, are required. The water-soluble antioxidants based on alkyl-substituted hydroxypyridines have been long used with success in medical practice. In the experiments with yeast cells, *S. cerevisiae*, we have revealed that 3-hydroxy-6-methyl-2-ethylpyridine essentially improves post-radiation recovery and raises survivability of the cells after γ-irradiation (^{60}Co, 800 Gy). Of special interest, can be some magnetic isotopes. Among three stable isotopes of magnesium, ^{24}Mg, ^{25}Mg, and ^{26}Mg with natural abundance approximately 79, 10, and 11 %, only ^{25}Mg has the nuclear spin (I = 5/2) and, hence, the nuclear magnetic field, while ^{24}Mg and ^{26}Mg have no nuclear spin (I = 0) and magnetic field. We have revealed that the rate constant of post-radiation recovery of cells after short-wave UV irradiation was twice higher for the cells enriched with magnetic ^{25}Mg, when compared to the cells enriched with the nonmagnetic isotope.

---

* E-mail address: koltover@icp.ac.ru.

Thus, the stable magnetic isotope of magnesium, as well as the non-toxic antioxidants, hold promises for creating novel radio-protectors suitable as nutrients for use at chronic radiation. [Supported by Russian Foundation for Basic Research, grant 10-03-01203a].

Keywords: Radiation stress, radiation protectors, antioxidant prophylaxis, stable magnetic isotopes, reliability, robustness

Introduction

Since the middle of last century, atomic power engineering has become a part of our everyday lives. Atomic catastrophes, like the catastrophes in Windscale (England, 1957), Three Mile Island (USA, 1979), Chernobyl (Ukraine, 1986), and Fukushima (Japan, 2011) may result from this kind of human being activity. Therefore, special precautions should be taken to reduce the harmful effects of radiation for specialists in atomic industry as well as for people in the contaminated territories. For emergency use, in the event of acute radiation, there is a number of effective radio-protectors based on the biologically active amines and aminothiols. However, these compounds are rather toxic. To defend people in case of chronic low-does radiation, alternative drugs, which would be non-toxic and suited to long-time applications as nutrients, are required. In this regard, of interest are natural and synthetic antioxidants, i.e., the phenolic compounds those valence-saturated molecules intercept active free radicals, thereby inhibiting free radical chain reactions. Of interest can be also some stable magnetic isotopes, especially, the magnetic isotope of magnesium, ^{25}Mg. The present article is a mini-review of our works on searching in this direction. It has been found that some non-toxic antioxidants, as well as, the stable magnetic isotope of magnesium, ^{25}Mg, hold promises for creating novel radio-protectors suitable as nutrients for use at chronic radiation.

Antioxidant Defense from Radiation Damage

By the middle 50s of the last century, it was already known that free radicals, including solvated electrons and hydroxyl radicals ($OH^{\bullet}$), arise in water under the action of ionizing radiation and such products of free radical oxidation, like radiation-induced peroxides of unsaturated fatty acids, oxidized nucleotides in DNA and oxidized proteins, appear in cells and animal tissues on exposure to ionizing radiation (see [1-5] and references therein). At the same years, in the former Soviet Union as well as behind the former "iron curtain" it has been found that antioxidants can essentially reduce the harmful effects of radiation. Some of them, if timely administered, can protect animals even against lethal doses of irradiation.

Antioxidants, by definition, are substances capable of terminating branching free-radical chain oxidation. These are mainly derivatives of secondary aromatic amines and phenols and organic phosphites and sulfides, those valence-saturated molecules containing an active hydrogen atom can react with active free radicals. A relatively unreactive free radical of the antioxidant, thus formed, cannot participate in chain propagation reactions and is destroyed upon collision with another radical or the vessel wall [6].

The most remarkable radioprotection effects were found with some aminothiols, the compounds of general formula $R_1R_2N(CH_2)_nSX$, where R_1 and R_2 are hydrogen atoms, alkyl or amino-alkyl radicals, X is acid residue $H_2PO_3^-$ or HSO_3^-, $n = 2$ or 3. For example, there are known the experiments with mice and rats when the intraperitoneal injections of 2-mercaptoethylamine (cysteamine) 5-15 min prior to the lethal-dose irradiation have prevented death of 80-100 % of the irradiated animals [2, 3]. Another aminothiol drugs, S-2-(aminopropylamino)-ethyl-phosphorothioate (Amifostine) and N-(2-mercaptoethyl)-1,3-diaminopropanel (WR-1065), have been reputed to protect against radiation lethality and toxicity caused by radiation therapy [7, 8]. There is the evidence for a preventive antioxidant action of the thiol-containing agents of this kind by nuclear transcription factor κB -mediated induction of the antioxidant enzyme, manganese superoxide dismutase in the cells [9]. Besides, a number of synthetic and natural inductors of cytokines, like β-interleukin, have been suggested as the effective protectors at radiation accidents [10].

However, these compounds are rather toxic and, for this reason, of little avail to defense against chronic low-dose irradiation. In the other words, they are of limited utility as prophylactic agents for workmen in atomic industry or people on the territories contaminated with radioactive fallouts. It has stimulated a search for radioprotection effects of natural compounds, such as caffeine, various flavonoids, ginseng, propolis from honeybee hives, geptrong from pure defermentated honey, and so on [11-15].

In 70s of the last century, the novel water-soluble antioxidants, based on alkyl-substituted hydroxypyridine, were synthesized in Russia, Institute of Chemical Physics of the former USSR Academy of Sciences. At the present time, 3-hydroxy-6-methyl-2-ethylpyridine succinate ('Mexidol') is used to treat brain circulation disorders, and hydrochloride of the same pyridine derivative ('Emoxipine') is used in ophthalmology. The same synthetic antioxidants proved to be effective geroprotectors, i. e., compounds that extend the life span of laboratory animals when added to food or drinking water on a regular basis. For example, additions of Emoxipine (3-hydroxy-6-methyl-2-ethylpyridine hydrochloride) to drinking water extended the average life spans of flies and mice by 24 and 38%, respectively (see, for example, reviews [6, 16, 17]). At the same period of time, it has been demonstrated that the same water-soluble non-toxic antioxidant may serve as the effective protector from radiation damage [17].

Of special importance were searches for the radioprotection effects in the processes of recovery of cells from radiation injuries. The survival of cells and multi-cellular organisms depends on the functioning of the enzyme systems that protect sub-cellular structures and cells from spontaneous and induced injuries, especially on the proper operation of the enzyme systems of reparation of genetic structures. Evidently, any factor capable to influence on efficiency and reliability of cell nanoreactors shows up itself most vividly under the unfavorable drastic conditions of post-radiation recovery. In fact, an exposure of a biological object to ionizing radiation followed by post-radiation recovery of cells and tissues may serve as analogue to the accelerated reliability testing in engineering when reliability of a device is tested under increased loads [18].

There have been undertaken the investigations of how the antioxidant Emoxipine influences on the post-radiation recovery of γ-irradiated yeast cells, *Saccharomyces cerevisiae*, the commonly accepted cell model in radiation biology. The yeast cells grown on the standard agar were suspended in sterile tap water, as the nutrient-free ("fasting") media, and exposed to γ-rays of ^{60}Co (800 Gy). To determine the kinetics of recovery we incubated the irradiated cells at 30°C in the same water, plated them out periodically on standard nutrient medium, and determined the survival by the standard technique of counting the macro-colonies (colony-forming units, CFU) [19].

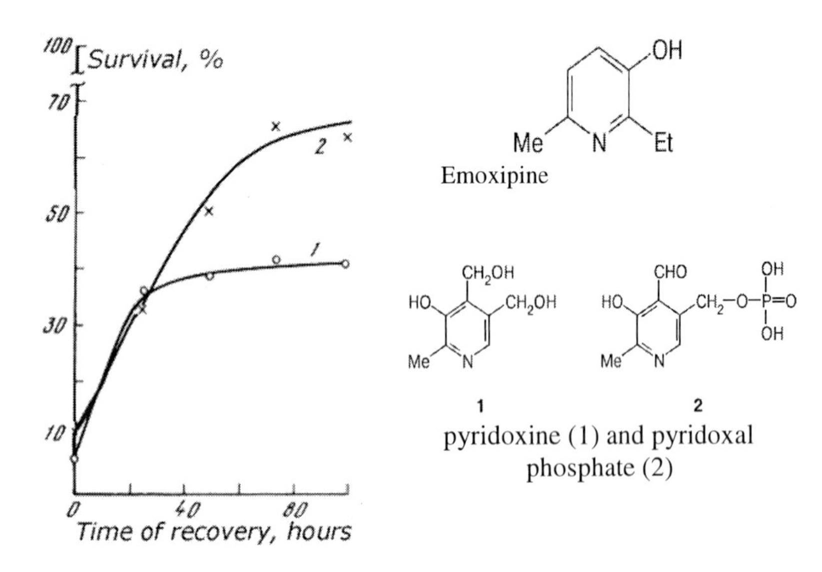

Figure 1. The influence of the water soluble antioxidant 3-hydroxy-6-methyl-2-ethylpyridine hydrochloride (Emoxipine) on kinetics of post-radiation recovery of the γ-irradiated yeast cells, *S. cerevisiae* (diploid strain Megri 139-B). The cells were irradiated by γ-rays (^{60}Co, 800 Gy, dose rate 0.028 Gy/s). Survival of the cells was estimated as their ability to form colonies on the nutrition agar: 1 - recovery in the control (without the antioxidant); 2 – recovery in the medium containing 10^{-7} M of the antioxidant added before irradiation [19].

The survival of cells transferred to agar immediately after irradiation was not more than a few percent (see Figure 1). In this case the injured genetic structures in most of the cells could not be repaired before the onset of mitosis and nonviable daughter cells were produced. Incubation in the nutrient-free water, where the cells do not divide, provides them with sufficient time for repair processes and leads to a corresponding increase in their survival.

One can see that, in the presence of Emoxipine added to the cells before the irradiation, the plateau on the survival curves is raised. There is the increase in the maximum survival values, up to 60% in reference to the control, which indicates a reduction in the proportion of irreversible (irreparable) radiation injuries in the presence of the antioxidant. These findings suggest that Emoxipine is really the promising prophylactic agent against radiation.

As for mechanisms of the beneficial effects of antioxidants, they now seems not so straightforward as it was in the last century. The results of analysis of the rate constants and actual concentrations of antioxidants raise doubts in the fact that antioxidants perform *in vivo* in as simple way as *in vitro*, i.e., only as free-radical inhibitors.

For example, in cells and tissues of aerobic organisms, the oxygen radical anion (superoxide radical, $O_2^{\bullet-}$) is formed that that serves the important source of chemically reactive ('toxic') oxygen species. However, there is also a specific enzyme, superoxide dismutase (SOD), that catalyzes dismutation of $O_2^{\bullet-}$ into hydrogen peroxide H_2O_2 and oxygen.

This enzyme, be it mitochondrial (Mn-SOD), cytosol (Cu,Zn-SOD) or periplasmatic bacterial enzyme (Fe-SOD), reacts with the $O_2^{\bullet-}$ radical anion with the rate constant ≈$2x10^9$ L mol^{-1} s^{-1}, while the rate constants for the reactions of the hydroxypyridine antioxidants with $O_2^{\bullet-}$ are no more than 10^2 L mol^{-1} s^{-1} [6]. Of course, since SOD protection has limited reliability, there exists a finite probability that O_2^- would penetrate the SOD barrier and then react with H_2O_2 to give hydroxyl radical ($OH^\bullet$). However, it is known that specialized enzymes, catalase and/or glutathione peroxidase, which catalyze hydrogen peroxide decomposition to water and oxygen, always occur near SOD [6].

It is generally accepted that hydroxyl radicals arise in cells and tissues under the action of ionizing radiation. However, $OH^\bullet$ is known to react with any organic molecules as a strong oxidant with rate constants close to the diffusion limit (>10^{10} L mol^{-1} s^{-1}). Therefore, none of antioxidants can compete for hydroxyl radicals *in vivo* with other organic molecules, which are obviously always present around the $OH^\bullet$ radicals in considerably greater numbers than molecules of any antioxidant.

Thus, the beneficial effects of antioxidants *in vivo* can hardly be interpreted on the basis of simple chemical analogy with the action of the same antioxidants as radical scavengers *in vitro* (see review [6] and the references therein). However, antioxidants, being unable to effectively scavenge free radicals *in vivo*, can increase the systems reliability through the prophylactic maintenance against reactive forms of oxygen. The particular protection mechanisms seem to be different for antioxidants of different types. For example, 2,6-di-*tert*-butyl-4-methylphenol ('butylated hydroxytoluene', BHT) prevents generation of the $O_2^{\bullet-}$ radicals as the by-products of electron transport in the mitochondrial nanoreactors [20]. Amifostine, WR-1065 and 'natural compounds' like flavonoids probably exert their radioprotection effects by inducing the expression of the SOD and other antioxidant enzymes [9]. As for the hydroxypyridine-based antioxidants, like Emoxipine, they are the analogs of pyridoxine and pyridoxal phosphate (see Figure 1).

As the anti-metabolites, they can inhibit the key enzymes of synthesis of amino acids and nucleotides, for example, glutamate-aspartate aminotransferase and RNA polymerases. Such the inhibition certainly retards mitosis and, thereby, provides the irradiated cells with additional time for restoring the genetic structures damaged by ionizing radiation [6].

It has been well documented that the beneficial effects of antioxidant prophylaxis, including the capability of some antioxidants to increase the life-spans of laboratory animals, are mediated via endocrine hormonal mechanisms, the hypothalamic-pituitary-adrenal axis and the hypothalamic-pituitary-thyroid axis [20, 21].

Furthermore, the special regulatory proteins those genes are sensitive to the changes in the redox state of intracellular medium have been recently found in human and animal organisms. For example, Sirt1 produces the epigenetic structural changes in chromatin which activate synthesis of protective proteins, among them – the antioxidant enzymes. Meanwhile, the redox state is controlled by the couples like GSH/GSSG, NADPH/$NADP^+$ and NADH/NAD^+ where the ratios of the reduced and oxidized components are dependent on the available concentrations of $O_2^{\bullet-}$ and antioxidants. Thus, more and more experimental results indicate that molecular mechanisms of the "antioxidant prophylaxis" are the tasks of the molecular systems biology.

STABLE MAGNETIC ISOTOPE OF MAGNESIUM-25 AS A POTENTIAL REMEDY AGAINST RADIATION DAMAGE

For preventive maintenance against radiation stress, some stable magnetic isotopes can be of interest as the basis of novel anti-stress medicine. There is a great variety of chemical elements which have both kinds of stable isotopes, nonmagnetic and magnetic, among them – carbon, oxygen, magnesium, calcium, iron, zinc, etc. In this regard, magnesium is of particular interest. As the most abundant intracellular divalent cation, Mg^{2+} is essential to regulate numerous cellular functions and enzymes, including ATP-synthase as the primary producer of ATP in mitochondria, chloroplasts, bacteria and archaea [22].

Among three stable magnesium isotopes, ^{24}Mg, ^{25}Mg and ^{26}Mg with natural abundance approximately 79, 10 and 11 %, only ^{25}Mg has nuclear spin (I = 5/2) that produces the magnetic field. Two other isotopes are spinless (I=0) and, hence, produce no magnetic fields [23]. In experiments with *Escherichia coli*, the commonly accepted cell model, the differences have been recently revealed in the quantitative parameters of the bacterial growth on the media supplied with different isotopes of magnesium [24]. Besides, the cells enriched with the magnetic ^{25}Mg, by comparison to the cells enriched with the nonmagnetic isotope, have demonstrated essentially higher viability estimated by counting colony forming units on the solid nutrient media [24].

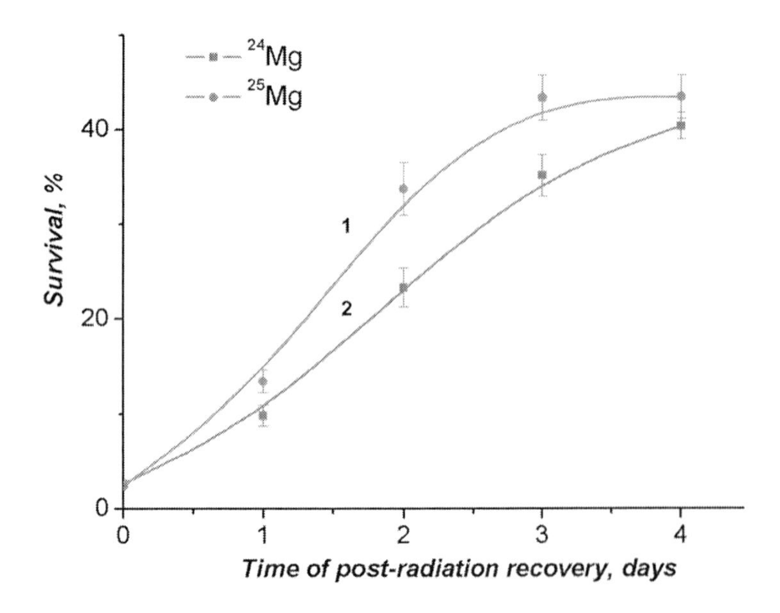

Figure 2. The difference in effects of magnetic and nonmagnetic isotopes of magnesium on post-radiation recovery of *S. cerevisiae*, diploid yeast cells. The cells enriched with magnetic ^{25}Mg or nonmagnetic ^{24}Mg were irradiated by the short-wave UV light (λ = 240-260 nm, dose ≈190 J/m^2). Survival of cells was estimated as their ability to form colonies on the nutrition agar: 1 - recovery of the cells enriched with ^{25}Mg; 2 – recovery of the cells enriched with ^{24}Mg [25].

We have undertaken the investigation of effects of magnetic and nonmagnetic isotopes of magnesium on post-radiation recovery of *S. cerevisiae* [25]. The yeast cells (diploid strain) were cultivated on the standard nutrient liquid media M3 supplemented with $^{24}MgSO_4$ or $^{25}MgSO_4$. After three days of the cultivation under aerobic conditions at 30^{0}C, the cells were washed from the nutrient liquid and suspended in nutrient-free ('fasting') media, i.e., sterile phosphate buffer, pH 7.0. Then the cells were irradiated by the short-wave ultraviolet light, whereupon they were left in the nutrient-free water at 30 ^{0}C (with shaking) to study kinetics of the post-radiation recovery. For this kinetics study, the aliquots of the cells were periodically seeded on the nutrient agar and the cell survival was monitored by the standard CFU technique.

Figure 2 represents the kinetics of recovery of yeast cells from radiation injuries. The survival of the cells transferred to agar immediately after irradiation was no more than a few percent. In this case the injured genetic structures in most of the cells could not be repaired before the onset of mitosis and nonviable daughter cells were produced. Incubation in the nutrient-free water, in which the cells do not divide, provides sufficient time for repair processes and leads to a corresponding increase in survival. From the kinetics curves represented on this Figure, we notice that the cells enriched with the magnetic isotope of magnesium, ^{25}Mg, are recovered essentially more effectively than the cell enriched with the nonmagnetic ^{24}Mg.

The kinetics of recovery of yeast cells from radiation injuries may be described by a function representing the reduction of the effective radiation dose, D_{eff}, with time:

$$D_{eff}(t) = D_0[k + (1-k)\exp(-\beta t)]$$

where D_0 is the radiation dose, t is time of post-radiation recovery in nutrient-free water, β is recovery rate constant, and k is fraction of irreversible injuries (model of A. Novick and L. Szilard [26]).

Table 1. Effect of magnetic 25Mg isotope on postradiation recovery of S. cerevisiae, diploid yeast cells, after short wave UV irradiation

	β, h^{-1}	k
^{24}Mg	0.032 ± 0.003	0.70 ± 0.14
^{25}Mg	0.058 ± 0.004*	0.61 ± 0.12

The table presents the results (m ± sd) of two independent experimental sessions with three samples simultaneously tested for every kind of the isotope in the same experimental succession (N = 3). *Difference between the means is significant at P = 0.02 [25].

Table 1 represents values of the kinetics parameters resulting from these experiments. While the fraction of irreparable injuries remained almost the same, the value of the rate constant of the post-radiation recovery was twice higher for the cells enriched with ^{25}Mg than for the cells enriched with ^{24}Mg. This is decisive evidence that the magnetic isotope of magnesium essentially more effectively promotes recovery of cells from radiation damages. Thus, we have, for the first time, documented the magnetic-isotope effect in radiation biology [25].

In addition, we have undertaken the investigation of effects of the magnetic and nonmagnetic isotopes of magnesium on post-radiation recovery of the haploid strain of *S. cerevisiae*. The enzyme system for reparation of DNA in cells of this auxotrophic strain is powerful enough to provide them the rather high resistance to radiation, so that there is no necessity to put the cells in the 'fasting' media [27]. The cells enriched with magnetic ^{25}Mg or nonmagnetic ^{24}Mg were irradiated by the short-wave UV light (λ = 240-260 nm) with dose 80 J/m^2 or 160 J/m^2. Survival of the cells was estimated as their ability to form colonies on the nutrition agar, CFU/ml, as a percentage to the non-irradiated control cells enriched with the same isotope, ^{25}Mg or ^{24}Mg, correspondingly.

From the data, presented on Figure 3, one can see that the cells enriched with ^{25}Mg manifest essentially greater resistance to the radiation damage, almost twice as much in the case of the higher radiation dose, when compared to the cells enriched with the nonmagnetic ^{24}Mg,. These findings suggest that the magnetic isotope of magnesium is indeed a remedy against radiation damage.

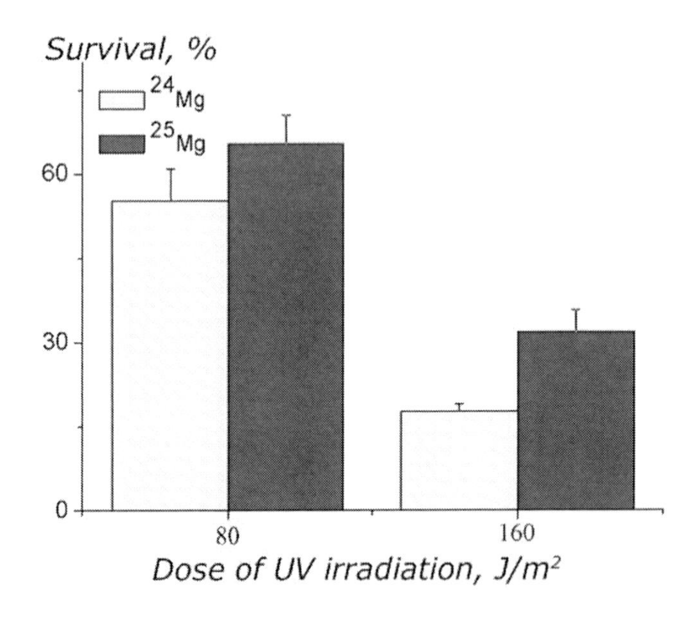

Figure 3. The difference in effects of magnetic and nonmagnetic isotopes of magnesium on survival of *S. cerevisiae*, haploid yeast cells (hsm3-1 rad2-1ade2-ura3), irradiated by UV light. The diagram presents the means ($m \pm sd$) of two independent experiments with three samples simultaneously tested for every kind of the isotope ($N = 3$).

One might suggest that the observed effects were caused by different levels of impurities in the growth media complemented with different isotopes of magnesium. However, it could hardly be the case. First, according to the mass-spectrometry data, amounts of contaminant elements in the stock solutions of the isotopes did not exceed 20-30 ppm. Second, amounts of the contaminants that were administered in the liquid growth media with glucose and other basic components have significantly exceeded amounts of the same contaminants administered with much less additions of the isotope stock solutions. Besides, the impurities that were administered into the growth media from the basic components, as well as the element contents of the solid nutrient agar media, were obviously the same in all experiments, independently of the magnesium isotopes. Hence, one can disregard impurities as a possible reason of higher efficiency of the media with magnetic ^{25}Mg than that of the media with the nonmagnetic one. It is apparent that the cells in the above cited experiments perceive the difference between magnetic and non-magnetic isotopes of magnesium, i. e. – they perceive the nuclear spin's magnetic field of^{25}Mg [24, 25].

In chemistry, the magnetic–isotope effect (MIE) is known for many elements, which have magnetic and nonmagnetic isotopes. MIE takes its origin from the law of conservation of electron angular moment (spin) in chemical reactions. It is agreed that MIE indicates that a pair of paramagnetic particles, free radicals or ion-radicals, participates in the reaction [29]. In biochemistry, MIE has been recently discovered in the experiments with mitochondria isolated from rat hearts. Namely, it was found that oxidative phosphorylation proceeds more effectively with magnetic isotope, ^{25}Mg, than with nonmagnetic ^{24}Mg or ^{26}Mg [29]. To explain MIE in the reaction of the ATP synthesis, it was suggested that the electron transfer from the anion phosphate group of ADP to Mg^{2+}–cation with formation of the virtual ion-radical pair, Mg^{+}–phosphate radical, takes place in the active center of ATP–synthase. Owing to the hyperfine interaction, the nuclear magnetic moment of ^{25}Mg changes the electron angular moment of this pair, converting it from the initial singlet state (total electron spin, $S = 0$) into the relatively long-lived triplet state ($S = 1$) in which the yield of the reaction of the ATP synthesis correspondingly increases [29]. In the light of the modern knowledge about mechanisms of oxidative phosphorylation the proposed mechanism of the MIE seems to be realistic but should certainly be proved experimentally [29, 30].

The similar MIE of ^{25}Mg is assumed to be in our experiments. Indeed, a large variety of biosynthesis is required for recovery of cells from radiation injuries. ATP as the main source of energy in the cells is most likely to be the limiting substrate for the adaptation metabolic processes. Of course, the Mg^{2+} ions perform not only the cofactor functions in synthesis and hydrolysis of ATP.

In addition, they have the impact on the structure-functional properties of RNA, RNA-polymerase, ribonuclease, and so on. Besides, there are the specialized proteins which regulate homeostasis and transport of Mg^{2+} in living cells [22].

Moreover, a novel role for Mg^{2+} as an intracellular second messenger has been recently discovered [28]. Up to date, however, there have not been findings of MIE except for the enzyme synthesis of ATP [29, 30]. Therefore, it is reasonable to believe that the kinetics of post-radiation recovery is limited by the spin-selective synthesis of ATP as the "bottleneck".

The post-radiation recovery proceeds with higher rate when the cell nanoreactors run on the magnetic isotope of magnesium, because the nuclear spin of ^{25}Mg catalyzes the ATP synthesis, hereby supplying the cells with more amount of ATP.

Conclusion

To defend people in case of chronic radiation, novel radiation protectors are required which would be suited to long-time applications as nutrients. Our experimental data with Emoxipine give grounds to expect that the non-toxic water soluble antioxidants, based on the alkyl-substituted hydroxypyridines are promising in this biomedicine field. Besides, the beneficial magnetic-isotope effects of ^{25}Mg *in vivo* have been documented.

Although the detailed mechanisms of the ability of the living cell to perceive magnetic properties of the atomic nuclei require further investigations, the discovery of the magnetic-isotope effect in radiation biology also holds promises for creating the novel radio-protectors suitable as nutrients for use at chronic radiation.

ACKNOWLEDGMENT

We are grateful to T. A. Evstykhina (Saint-Petersburg Institute of Nuclear Physics, Gatchina, Leningrad Region) her excellent assistance and D. M. Grodzinsky (Institute of Cell Biology and Genetic Engineering, Ukraine Academy of Sciences, Kyiv, Ukraine) for the fruitful discussions. We greatly acknowledge Dr. V. K. Karandashev, I. R. Moskvina and S. V. Nosenko (Institute of Microelectronics Technology and High Purified Materials, RAS, Chernogolovka), for the excellent element analysis of the solutions and media by the inductively coupled plasma mass-spectrometry (spectrometer ICP–MS "Element–2," USA). This work has been supported by Russian Foundation for Basic Research, project no. 10-03-01203a.

REFERENCES

[1] Tarusov, B. N. *Foundations of the Biological Action of Radioactive Radiation*, Medgiz, Moscow, 1954, 140 pp. (in Russian).

[2] Bacq, Z. M. *Chemical protection against ionizing radiation*, Thomas, Springfield, Ill. (USA), 1965, 328 p.

[3] Livesey, J. C.; Reed, D. J.; Adamson, L. F. *Radiation-protective drugs and their reaction mechanisms*, Noyes Publications, Park Ridge, N. J. (U.S.A.), 1985. 146 p.

[4] Kutlakhmedov, Y. A.; Korogodin, V. I.; and Koltover, V. K. *Foundations of Radiation Ecology*, Vyshcha shkola, Kiev (Ukraine), 2003, 320 pp. (in Ukrainian).

[5] Kutlakmedov, Y.; Voytizky, V.; Khuznyak, S. *Radiobiology*, Kiev University Publisher's, Kiev (Ukraine), 2011, 542 pp. (in Ukrainian).

[6] Koltover, V. K. Antioxidant biomedicine: from free radical chemistry to systems biology mechanisms. *Russian Chem. Bulletin. Inter. Edition*, 2010, 59 (1), 37-42.

[7] Pathak, U.; Raza, S. K.; Kumar, P.; Vijayaraghavan, R.; Jaiswal, D. Amifostine: An effective prophylactic agent against sulphur mustard toxicity. *Def. Sci. J.*, 2002, 52, 439-444.

[8] Azzam, E. I.; Jay-Gerin, J.-P.; Pain, D. Ionizing radiation-induced metabolic oxidative stress and prolonged cell injury. *Cancer Lett.*, 2012,

[9] Murley, J. S.; Kataoka, Y.; Weydert, Ch. J.; Oberley, L. W.; Grdina, D. J. *Free Radical Biol. and Med.*, 2006, 40 (6), 1004-1016.

[10] Baranov, A. E.; Rozhdestvenskiy, L. M. *Radiation Biology. Radiation Ecology*, 2008, 48 (3), 287-302 (in Russian).

[11] Lee, T. K.; Johnke, R. M.; Allison, R. R.; O'Brien, K. F.; Dobbs, L. J. Radioprotective potential of ginseng. *Mutagenesis*, 2005, 20 (4), 237-243.

[12] Montoro, A.; Almonacid, M.; Serrano, J.; Saiz, M.; Barquinero, J. F.; Barrios, L.; Verdu, G.; Perez, J.; Villaescusa, J. I. Assessment by cytogenetic analysis of the radioprotection properties of propolis extract. *Radiat. Prot. Dosim.*, 2005, 115 (1-4), Part 1., Sp. Iss. 461-464.

[13] Rithidech, K. N.; Tungjai, M.; Whorton, E. B. Protective effect of apigenin on radiation-induced chromosomal damage in human lymphocytes. *Mutat. Res. Genet. Toxicol. Environ. Mutagen.*, 2005, 585 (1-2), 96-104.

[14] Mauryam, D. K.; Devasagayam, T. P. A.; Nair, C. K. K. Some novel approaches for radioprotection and the beneficial effect of natural products. *Indian J. Exp. Biol.*, 2006, 44 (2), 93-114.

[15] Kovaltsova, S. V.; Fedorova, I. V.; Gracheva, L. M.; Mashistov, S. A.; Korolev, V. G. The Geptrong pharmaceutical product increases efficiency of postreplication repair of permutation intermediates in yeast *Saccharomyces cerevisiae. Russ. J. Genet.*, 2008, 44 (11), 1272-1279.

[16] Obukhova, L. K.; Emanuel, N. M. The role of free radical reactions of oxidation in molecular mechanisms of aging of living organisms. *Russian Advances in Chemistry*, 1983, 52, 353-372.

[17] Burlakova, E. B. Bioantioxidants. *Mendeleev. Chem. J. (Engl. Transl.)*, 2007, 51, 3-12.

[18] Grodzinsky, D. M.; Vojtenko, V. P.; Kutlakhmedov, Y. A.; Koltover, V. K. *Reliability and Aging of Biological Systems,* Naukova Dumka, Kiev (fSU), 1987, 172 pp. (in Russian).

[19] Kol'tover, V. K.; Kutlakhmedov, Y. A.; Afanaseva, E. L. Recovery of cells from radiation-induced damages in the presence of antioxidants and the reliability of biological systems. *Doklady Biophysics*, 1980, 254, 159-161.

[20] Koltover, V. K. Bioantioxidants: The systems reliability standpoint. *Toxicology and Industrial Health*, 2009, 25 (4-5), 295-299.

[21] Goncharova, N. D.; Shmaliy, A. V.; Bogatyrenko, T. N.; Koltover, V. K. Correlation between the activity of antioxidant enzymes and circadian rhythms of corticosteroids in *Macaca mulatta* monkeys of different age. *Exp. Gerontol.*, 2006, 41, 778-783.

[22] Romani, A. M. P. Cellular magnesium homeostasis. *Arch. Biochem. Biophys.*, 2011, 512, 1-23.

[23] Grant, D. M.; Harris, R. K. (Eds.). *Encyclopedia of Nuclear Magnetic Resonance*, Wiley, Chichester, 1996.

[24] Koltover, V. K.; Shevchenko, U. G.; Avdeeva, L. V.; Royba, E. A.; Berdinsky, V. L.; Kudryashova, E. A. Magnetic isotope effect of magnesium in the living cell. *Doklady Biochemistry and Biophysics*, 2012, 442 (1-2), 12-14.

[25] Grodzinsky, D. M.; Evstyukhina, T. A.; Koltover, V. K.; Korolev, V. G.; Kutlakhmedov, Y. A. Effect of the magnetic isotope of magnesium, ^{25}Mg, on post-radiation recovery of *Saccharomyces cerevisiae. Dopovidi NAN Ukraiiny (Reports Nat. Acad. Sci. Ukraine)*, 2011,. No. 12, 153-157 (in Ukrainian).

[26] Novick, A.; Szilard, L. Experiments on light-reactivation of ultra-violet inactivated bacteria. *Proc. Natl. Acad. Sci. USA*, 1949, 35, 591-600.

[27] Korolev, V. G. Excision reparation of the damaged bases of DNA. AP-endonucleases and DNA polymerases. *Russ. J. Genet.*, 2005, 41 (10), 1301-1309.

[28] Li, F. Y.; Chaigne-Delalande, B.; Kanellopoulou, C.; Davis, J. C. H. F.; Douek, M. D. C.; Cohen, J. I.; Uzel, G.; Su, H. C.; Lenardo, M. J. Second messenger role for Mg^{2+} revealed by human T-cell immunodeficiency. *Nature*, 2011, 475, 471-476.

[29] Buchachenko, A. L. *Magnetic Isotope Effect in Chemistry and Biochemistry*, Nova Science Publishing, New York, 2009.

[30] Koltover, V. K. Nuclear spin catalysis in nanoreactors of living cells. *Biophysics (Moscow)*, 2012, 57, No. 6 (in press).

In: Ionizing Radiation ISBN: 978-1-62257-343-1
Editors: Eduard Belotserkovsky and Ziven Ostaltsov

Chapter 7

IONIZING RADIATION: APPLICATIONS, SOURCES AND BIOLOGICAL EFFECTS

Carlos Alexandre Fedrigo[1], Ivana Grivicich[2] and Adriana Brondani da Rocha[3]

[1]Programa de Pós-Graduação em Medicina e Ciências da Saúde, Pontifícia Universidade Católica do Rio Grande do Sul, Porto Alegre, RS, Brasil
[2]Programa de Pós-Graduação em Biologia Celular e Molecular Aplicada à Saúde, Universidade Luterana do Brasil, Canoas, RS, Brasil
[3]Programa de Pós-Graduação em Genética e Toxicologia Aplicada, Universidade Luterana do Brasil, Canoas, RS, Brasil; Conselho de Informações sobre Biotecnologia, São Paulo, SP, Brasil

ABSTRACT

Ionizing radiation (IR) occurs naturally at low doses. Mostly, when IR induces DNA damage, our cells are able to repair the error. These damages can be direct, causing ionization - ejection of electrons from molecules - of DNA atoms, or indirect when there is an interaction with water and other cellular molecules, leading to a generation of reactive oxygen species.

When the cells are not able to fix the error, a sequential multistep process generates several types of chemically stable lesions, which can lead to cell death. Therefore, IR is being used for cancer treatment, reaching 50% of all tumor types. New techniques and combined therapies with radiosensitizers and chemotherapics are being evaluated in order to improve the efficacy of IR treatment. In fact, significant progress has been made in our understanding of the basic mechanisms of radiation injury and its cellular processing in both normal and malignant cells. New discoveries offer increasing opportunities for clinical applications. Our increasing understanding of the basic mechanisms controlling the cell cycle and apoptosis provides important molecular markers for the caracterization of injury responses in different types of normal and malignant cells, and important molecular targets for future therapeutic intervention. This chapter discusses some of these issues in the context of their relevance to the clinical effects of radiation on tumor and normal cells.

A detailed understanding of the mechanisms and pathways of radiation injury and repair may lead to the design of new biologic or chemical response modifiers to improve the therapeutic ratio of radiation treatments in human cancer.

1. Nature of Ionizing Radiation Lesions in Mammalian Cells

Radiation has a wide range of energies forms of the electromagnetic spectrum, and has two major divisions: radiation with lower energy levels, that is called non-ionizing and, radiation with higher energy levels that is called ionizing. The radiation that has enough energy to move atoms in a molecule around or cause them to vibrate, but not enough to remove electrons, is referred to as non-ionizing radiation. Examples of this kind of radiation are radio waves, visible light, microwaves, infrared light, heat waves, and visible portions of the spectrum into the ultraviolet light. On the other hand, the radiation that has enough energy to remove tightly bound electrons from atoms, thus creating ions which deposits energy in the cells of the tissues it passes through is called ionizing radiation (IR). IR occurs naturally at low doses, as example, X-rays, gamma rays and non-visible portions of the spectrum into the ultraviolet light [1].

The X-rays and gamma rays are widely used for radiotherapy treatments, presenting a good alternative for surgery on long-term control of more than 50% of all tumors. Besides the curative role of radiation therapy, several cancer patients gain valuable palliation of symptoms by radiation [2]. X- rays and gamma rays are sparsely ionizing radiations, considered low LET (linear energy transfer) electromagnetic rays and further composed of massless particles of energy are called photons. X-rays are generated by a device that excite electrons (e.g. cathode ray tubes and linear accelerators), while gamma rays originate from the decay of radioactive substances (e.g.cobalt-60, radium and cesium) [3].

In the time-scale effects of IR, three phases can be distinguished. The first and quickest, physical phase, is represented by the interaction between charged particles and the atoms, interacting mainly with orbital electrons and ejecting some of them – process called ionization – and raising others to higher levels of energy – process called excitation. The second phase, called chemical phase, is the stage where ionized or excited atoms react with other molecules, leading to disruptions of chemical bonds, breaking molecules and releasing free radicals. The last and longer, the biological phase, corresponds of the residual chemical damage, where the lesions that were not repaired eventually lead to cell death [4].

Considering that all photons directed to a tissue have sufficient energy to produce damage it is surprising that the majority of damage lays down on the ionizing capability, in other words, the reactions between an eject electron and other molecules. As the cells are composed mostly by water, irradiated water undergoes ionization and radiolysis, producing hydrated free electrons (e-) and extremely reactive oxygen species (ROS) as *OH. These products can attack the DNA, lipids and proteins within the cells. This is normally referred to as the oxygen-fixation hypothesis (Figure 1) [5].

High-energy radiation damages genetic material of cells and thus blocking their ability to divide and proliferate further [6]. On DNA, the nature of damage done can be different: single-strand breaks (SSBs), double-strand breaks (DSBs), DNA base damage and DNA protein crosslinks [7].

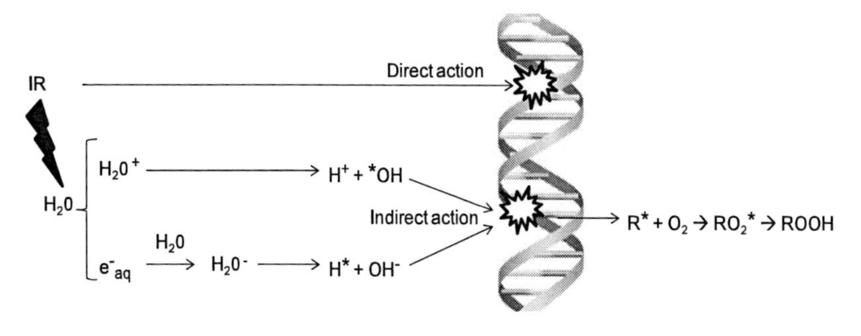

Figure 1. Reaction pathway resulting from water ionization and oxygen-fixation. The production of free radicals (R*) due to radiation (IR) generates highly reactive molecules that attacks DNA, initiating a chain of events that may result in biological damage. The unstable R* molecules will react with oxygen resulting in ROOH. At this stage the damage was "fixated" and participates on DNA damage responses (DDRs).

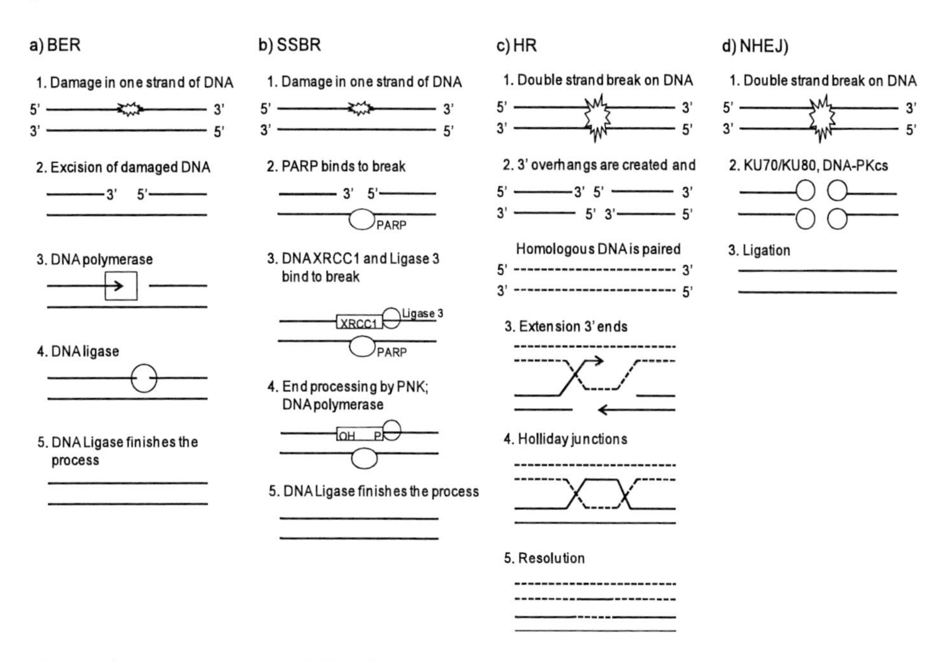

Figure 2. Repair system to correct DNA damages: a) base excision repair (BER), b) single-strand break repair (SSBR), c) homologous recombination (HR) and d) and non-homologous end-joining (NHEJ).

Double strand DNA breaks are irreparable and more responsible than the single strand DNA breaks for most of cell killing in cancer as well as surrounding normal cells [3]. Moreover, ROS can produce a large spectrum of lesion in cellular macromolecules, for example, lipid peroxidation, stimulating signal transductions pathways – including protein kinase C (PKC), JNK, ceramide and MAPK activation [7].

However, considering the obvious importance of the DNA, cells developed a complex system to activate pathways and responses to correct the mistakes during replication and damages produced by an external source, namely: a) base excision repair (BER), b) single-strand break repair (SSBR), c) homologous recombination (HR) – related to the DSBs –, d) and non-homologous end-joining (NHEJ) (Figure 2). Ultimately, if the error cannot be corrected and cell cannot be saved, it will lead to death.

2. Cellular Responses to Ionizing Radiation and Damage Processing

The DNA damage response (DDR) is not a single pathway, but a combination of interrelated reactions that will determine the result of a damage caused by radiation on DNA. These DNA damage sensors include: a) the ataxia telangiectasia mutated (ATM) protein, together with the MRN protein complex; b) DNA-dependent protein kinase repair complex (DNA-PKcs); c) the ataxia telangiectasia-related (ATR) kinase; and d) DNA mismatch repair (MMR) complex, among others [7].

As the IR progress the first cellular response is the recruitment of several different proteins to the site of damage, producing a cluster that can be visualized with proper staining. The region is commonly called ionizing radiation induced 'foci' (IRIF) [8]. One of the first of these events is the phosphorylation of the histone H2AX, leading to γH2AX (phosphorylated form), which can be measured in the DSBs formation areas [9].

The phosphorylation of H2AX is done primarily by the ATM protein, and needs the MRN protein complex to be fully activated, which is formed by three proteins: MRE11, RAD50 and NBS1. Inactive, but present on cells, ATM becomes active with a DSB originating a γH2AX that spreads through the chromatin to both directions; mediated by the additional protein MDC1 (acts as a mediator by directly binding to ATM and H2AX). In those cells where there is no ATM protein, the γH2AX is formed with the help of DNA-PKcs through a similar mechanism: unable to act as a damage sensor, it recruits the Ku70-Ku80 complex, which binds to the ends of DSBs. The binding of Ku70-Ku80 and DSBs will be the signal to DNA-PKcs to phosphorylate H2AX [2].

The third capable kinase to phosphorylate H2AX is ATR. This kinase, on the other hand, acts on SSB on DNA and stalled of broken replication forks, for example, stretches of single-stranded DNA done by the ATM-MRN complex. To be fully active, ATR forms a dimer with another protein, ATR interacting protein (ATRIP), and as previously, is the sensor protein that recruits the complex to the site of damage [10].

We can call attention to MMR and BER systems either. MMR is a postreplicative mechanism that ensures the application of the Watson-Crick base pairing principle on DNA double helix, discriminating mismatches resulting from DNA polymerase errors, and rectifying them to avoid mutation [11]. The MMR multiprotein system is formed by three subprocesses: mismatch recognition (MSH2/MSH6 or MSH2/MSH3), mismatch excision (binding of MLH1/PMS2 or MLH1/MLH3 to MutSa) and the nucleotides removal (exonuclease EXO1). Followed by DNA polymerase and DNA ligase, the damage caused by IR would be, eventually, fixed (12). Also a multistep pathway, BER complex reacts to the creation of abasic or apurinic/apyrimidinic sites [13]. Generally, BER corrects DNA base lesions due to oxidative, alkylation, deamination damages via two general pathways: for one nucleotide repair (short-patch BER) or for 2 to 15 nucleotides repair (long-patch BER) [13, 14]. BER is one of the main DNA repair system that reacts to damages induced by oxidative base modifications, for example from ROS, as well for DNA SSBs after IR [15, 16].

Finally, an important distinction must be made regarding HR and NHEJ, majors DSBs repair pathways [17]. Homologous recombination implies the use of an undamaged DNA as template to repair the DSB, providing a model of the base sequence around the error site to DNA polymerase. The process starts with the cut of the strands with an exonuclease – MRN complex is involved – and is followed by the action of replication protein A (RPA), common response of the cell to single-stranded regions on DNA. The central protein RAD51 takes place, searching for homologous DNA and strand invasion, and after this, the polymerase acts. In the end, enzymes called resolvases cut the DNA on crossover points and ligases finish the whole process. For NHEJ repair system, two DNA DSBs are joined together without the need of homologous DNA sequences. Despite being less precise – with small deletions or insertions – NHEJ is far quicker than HR. In brief, involve nucleases that remove the damage and polymerases to repair it. Important proteins involved here: the Ku70-Ku80 dimer, protecting the ends from degradation and recruiting DNA-PKcs; polynucleotide kinase (PNK), removing sugar groups from ends and leaving the base with 3' phosphate; and again ligases to finish the repair [18].

Besides being part of big repair machinery, all sensors described still contribute to the regulation of cell cycle checkpoints. The most well known are p53 and NF-kB. Activated, these transcription factors (TFs), control gene expression, cell cycle and apoptosis.

The tumor suppressor gene *p53* helps to maintain the genomic integrity and has an important role in cell fate: will the damaged cell survive and repair the DNA damage and present growth arrest, or will it activate programmed cell death? It is known that *p53* works mostly through direct binding to those sequences that exhibit a regulatory function in genes. However, a dual response seems to be present on *p53*, acting both on DNA repair and cell cycle arrest to overdue stresses, and alternatively, on apoptosis, leading the cell to death [19]. Ultimately these responses can affect the IR and chemotherapies approaches [20, 21]. Nevertheless, this is controversial, since some investigators have found that apoptosis after IR treatment can occur through alternative signaling pathways independent of *p53* status [22, 23].

The TFs family related NFkB is regulated by two kinases, IKKα and IKKβ, and commonly associated with chronic inflammation, cancer, neurodegenerative disorders, diabetes and stroke. In cancer is considered one of the major TFs implicated in growth, proliferation, invasion, inhibition of apoptosis and metastasis [24, 25]. The regulation of specific genes that expresses growth factors, for example, Cyclin D1, VEGF, matrix metalloproteinases, TNF-α, IL-1, IL-8 and cyclooxygenase-2 in response to radiotherapy is one of the main mechanisms to induce radioresistance [26, 27].

All this knowledge about DNA damage sensors, DDRs and gene expression is obtained through basic research on laboratory, commonly using cell lines derived from human tumors. The genetic differences between these cell lines, responses to IR and other treatments are of major importance to predict new modalities of study to improve current therapies and improve quality of life and survival of patients.

3. Use of Radiation Survival Parameters *In Vitro* As Predictors to Response to Radiotherapy

However, the main problem of predictive assays is the difficulty to transpose the tumor response from the bench to bedside, the so called translational research. Investigations to improve the outcome of radiotherapy through the study of biological parameters such as tumor hypoxia, tumor cell kinetics and radiosensitivity are of major importance. It has long been recognized that tumors are heterogeneous in their radiation response. In some cases, the degree of radiosensitivity was believed to be related to intrinsic properties (DNA repair potential and proliferation status) and to extrinsic properties (degree of hypoxia within the tumor) of the tumor cells, which impacted their capacity to withstand radiation insult [28].

Moreover, there is a large variation in normal tissue toxicity between different types of tumors, and even within the same tumor, patients responses are broad, which leads to the necessity of very specific and better treatment schedules for individual patients [29].

To quote some of the normal tissue toxicity problems faced: dose of radiation, type of fractionation and duration of treatment, age, life style, smoking, chemotherapeutics and, most important, genetic factors, ultimately related to the activation or inactivation of the cellular responses to the damages done by IR mentioned before [30].

The results seem controversial depending on the technique used. For example, with comet assay or micronucleus assay, the use of initial DNA damage as a clinical tool was not possible due to such different results. On the other hand, through flow cytometry analysis, just one in eight studies did not agree with the use of this technique to measure induced DNA damage in peripheral blood lymphocytes to predict normal tissue toxicity to radiation, controlled by the already mentioned ATM disease and epigenetic factors [29].

Even with the gold standard assay to measure cellular response to radiotherapy in laboratory, clonogenic cell survival assay [31], there are doubts. Determining *ex vivo* tumor survival fraction at 2 Gy of IR led authors to a correlation of predictive/prognostic with clinical outcome, while others have found none correlation [32]. Besides the contradictory results, technical problems present deterrent factors against this assay: not all cell lines fixate and generate colonies *in vitro*; those that generate may present a low plating efficiency; the several weeks time to get the results can be a problem at treatment decision time.

Recent experimental evidence indicates that many solid cancers have a hierarchical organization structure with a subpopulation of cancer stem cells (CSCs). The ability to identify CSCs prospectively now allows for testing the responses of CSCs to treatment modalities like radiotherapy [33]. If tumor growth after radiotherapy is a property of CSCs, the response of these cells to IR is a critical parameter for curability. For instance, a IR resistance of CD133+ cells was reported in gliomas, and this resistance was attributed to constitutive activation of the DNA repair checkpoints and inhibition of the corresponding kinase radiosensitized CD133+ cells [34]. Moreover, it was reported a radioresistance of breast CSCs but, contrary to glioma, CSCs breast produced less reactive oxygen species in response to IR indicating a high level of expression of free-radical scavengers [35].

Limiting normal tissue toxicity and radioresistant tumors are still linked to life-threatening radiation treatment failure. A novel approach is the search for radiosensitizers, which lower the radiation dose-response threshold for cancer cells without enhancing the radiosensitivity of normal cells [36]. The last years provided a lot of knowledge regarding the understanding of molecular pathways in cancer, some of them described above.

Allied with this, the availability of new agents for combined therapy to overcome radioresistance has made possible to improve therapy, and moreover, aim individual patients from specific therapies [37, 38]. Among these, the camptothecins analogs [39, 40] and inhibitors of histone deacetylases [41].

Since the vast use of radiotherapy in all types of cancers, the outcome of assays to predict response to IR would be of great impact, resulting in better radiotherapy protocols and in an improvement of individual response and prognosis.

CONCLUSION

The radiation that has enough energy to remove electrons from atoms, generating ions in which deposits energy in the tissues it passes through, is called IR. IR occurs naturally at low doses, as example: X-rays, gamma rays and non-visible portions of the spectrum into the ultraviolet light. The X-rays and gamma rays are widely used for radiotherapy as a cancer treatment. IR has been shown to induce various types of chromosomal DNA damages. Among these DNA damages, DNA double strand breaks (DSBs) are the most severe damages resulting in cell death or chromosome abnormalities. Proteins associated with DNA repair, such as phosphorylated form of histone H2AX, a histone variant of H2A, and a DNA recombinase RAD51, has been shown to form radiation-induced repair foci at sites containing DNA damage. For instance, the genetic differences between tumor cell types and their responses to IR are of major importance to improve current therapies and lead to increase the time of patients survival. Thus, the study of biological parameters such as tumor hypoxia, tumor cell kinetics and radiosensitivity are crucial to improve the outcome of radiotherapy.

REFERENCES

[1] Shrieve DC, Klish M, Wendland MM, Watson GA. Basic principles of radiobiology, radiotherapy, and radiosurgery. *Neurosurg. Clin. N. Am.* 2004; 15(4):467-79, x.

[2] Joiner M, Kogel Avd. Basic clinical radiobiology. 4th ed. London: Hodder Arnold; 2009.

[3] Baskar R, Lee KA, Yeo R, Yeoh KW. Cancer and radiation therapy: current advances and future directions. *Int .J. Med. Sci.* 2012;9(3):193-9.

[4] Boag JW. Twelfth Failla Memorial Lecture. *The time scale in radiobiology.* New York: Academic Press; 1975.

[5] Hall EJ, Giaccia AJ. *Radiobiology for the radiologist.* 6th ed. Philadelphia: Lippincott Williams and Wilkins; 2006.

[6] Jackson SP, Bartek J. The DNA-damage response in human biology and disease. *Nature.* 2009;461(7267):1071-8.

[7] Criswell T, Leskov K, Miyamoto S, Luo G, Boothman DA. Transcription factors activated in mammalian cells after clinically relevant doses of ionizing radiation. *Oncogene.* 2003;22(37):5813-27.

[8] Franken NA, ten Cate R, Krawczyk PM, Stap J, Haveman J, Aten J, et al. Comparison of RBE values of high-LET alpha-particles for the induction of DNA-DSBs, chromosome aberrations and cell reproductive death. *Radiation oncology.* 2011;6:64.

[9] Olive PL, Banath JP. Phosphorylation of histone H2AX as a measure of radiosensitivity. *International Journal of radiation oncology, biology, physics.* 2004;58(2):331-5.

[10] Shiloh Y. ATM and related protein kinases: safeguarding genome integrity. *Nature reviews Cancer.* 2003;3(3):155-68.

[11] Kunz C, Saito Y, Schar P. DNA Repair in mammalian cells: Mismatched repair: variations on a theme. *Cellular and molecular life sciences : CMLS.* 2009;66(6): 1021-38.

[12] Jiricny J. The multifaceted mismatch-repair system. *Nature reviews Molecular cell biology*. 2006;7(5):335-46.

[13] Hitomi K, Iwai S, Tainer JA. The intricate structural chemistry of base excision repair machinery: implications for DNA damage recognition, removal, and repair. *DNA repair*. 2007;6(4):410-28.

[14] Robertson AB, Klungland A, Rognes T, Leiros I. DNA repair in mammalian cells: Base excision repair: the long and short of it. *Cellular and molecular life sciences : CMLS*. 2009;66(6):981-93.

[15] David SS, O'Shea VL, Kundu S. Base-excision repair of oxidative DNA damage. *Nature*. 2007;447(7147):941-50.

[16] Helleday T, Petermann E, Lundin C, Hodgson B, Sharma RA. DNA repair pathways as targets for cancer therapy. *Nature reviews Cancer*. 2008;8(3):193-204.

[17] Hiom K. Coping with DNA double strand breaks. *DNA Repair (Amst)*. 2010; 9(12):1256-63.

[18] Hartlerode AJ, Scully R. Mechanisms of double-strand break repair in somatic mammalian cells. *Biochem. J.* 2009;423(2):157-68.

[19] Jackson JG, Post SM, Lozano G. Regulation of tissue- and stimulus-specific cell fate decisions by p53 in vivo. *J. Pathol.* 2011; 223(2): 127-36.

[20] Bertheau P, Espie M, Turpin E, Lehmann J, Plassa LF, Varna M, et al. TP53 status and response to chemotherapy in breast cancer. *Pathobiology*. 2008;75(2):132-9.

[21] Ventura A, Kirsch DG, McLaughlin ME, Tuveson DA, Grimm J, Lintault L, et al. Restoration of p53 function leads to tumour regression in vivo. *Nature*. 2007;445(7128):661-5.

[22] Stegh AH, Brennan C, Mahoney JA, Forloney KL, Jenq HT, Luciano JP, et al. Glioma oncoprotein Bcl2L12 inhibits the p53 tumor suppressor. *Genes Dev*. 2010;24(19): 2194-204.

[23] Fedrigo CA, Grivicich I, Schunemann DP, Chemale IM, Santos DD, Jacovas T, et al. Radioresistance of human glioma spheroids and expression of HSP70, p53 and EGFr. *Radiat. Oncol.* 2011;6:156.

[24] Kumar A, Takada Y, Boriek AM, Aggarwal BB. Nuclear factor-kappaB: its role in health and disease. *J. Mol. Med. (Berl)*. 2004;82(7):434-48.

[25] Naugler WE, Karin M. NF-kappaB and cancer-identifying targets and mechanisms. *Curr. Opin. Genet. Dev*. 2008;18(1):19-26.

[26] Deorukhkar A, Krishnan S. Targeting inflammatory pathways for tumor radiosensitization. *Biochem. Pharmacol*. 2010;80(12):1904-14.

[27] Tamatani T, Azuma M, Ashida Y, Motegi K, Takashima R, Harada K, et al. Enhanced radiosensitization and chemosensitization in NF-kappaB-suppressed human oral cancer cells via the inhibition of gamma-irradiation- and 5-FU-induced production of IL-6 and IL-8. *Int. J. Cancer*. 2004;108(6):912-21.

[28] Redmond KM, Wilson TR, Johnston PG, Longley DB. Resistance mechanisms to cancer chemotherapy. *Front Biosci*. 2008;13:5138-54.

[29] Henriquez-Hernandez LA, Bordon E, Pinar B, Lloret M, Rodriguez-Gallego C, Lara PC. Prediction of normal tissue toxicity as part of the individualized treatment with radiotherapy in oncology patients. *Surg. Oncol.* 2011.

[30] Azria D, Betz M, Bourgier C, Sozzi WJ, Ozsahin M. Identifying patients at risk for late radiation-induced toxicity. *Crit. Rev. Oncol. Hematol.* 2010.

[31] Franken NA, Rodermond HM, Stap J, Haveman J, van Bree C. Clonogenic assay of cells in vitro. *Nat. Protoc.* 2006;1(5):2315-9.

[32] Torres-Roca JF, Stevens CW. Predicting response to clinical radiotherapy: past, present, and future directions. *Cancer Control.* 2008; 15(2):151-6.

[33] Vlashi E, McBride WH, Pajonk F. Radiation responses of cancer stem cells. *Journal of cellular biochemistry.* 2009;108(2):339-42.

[34] Bao S, Wu Q, McLendon RE, Hao Y, Shi Q, Hjelmeland AB, et al. Glioma stem cells promote radioresistance by preferential activation of the DNA damage response. *Nature.* 2006;444(7120):756-60.

[35] Phillips TM, McBride WH, Pajonk F. The response of CD24(-/low)/CD44+ breast cancer-initiating cells to radiation. *J. Natl. Cancer Inst.* 2006;98(24):1777-85.

[36] Camphausen K, Tofilon PJ. Combining radiation and molecular targeting in cancer therapy. *Cancer biology and therapy.* 2004;3(3):247-50.

[37] Verheij M, Vens C, van Triest B. Novel therapeutics in combination with radiotherapy to improve cancer treatment: rationale, mechanisms of action and clinical perspective. *Drug Resist. Updat.* 2010;13(1-2):29-43.

[38] Zaidi SH, Huddart RA, Harrington KJ. Novel targeted radiosensitisers in cancer treatment. *Curr. Drug Discov. Technol.* 2009;6(2):103-34.

[39] Miura K, Sakata KI, Someya M, Matsumoto Y, Matsumoto H, Takahashi A, et al. The combination of olaparib and camptothecin for effective radiosensitization. *Radiat. Oncol.* 2012;7(1):62.

[40] Illum H. Irinotecan and radiosensitization in rectal cancer. *Anti-cancer drugs.* 2011;22(4):324-9.

[41] Konsoula Z, Velena A, Lee R, Dritschilo A, Jung M. Histone deacetylase inhibitor: antineoplastic agent and radiation modulator. *Adv. Exp. Med. Biol.* 2011;720:171-9.

In: Ionizing Radiation
Editors: Eduard Belotserkovsky and Ziven Ostaltsov
ISBN: 978-1-62257-343-1

Chapter 8

DOSIMETRY CHARACTERIZATION OF A PILOT-SCALE COBALT-60 γ-IRRADIATION FACILITY FOR THE RADIATION STERILISATION OF INSECTS

Khaled Farah[1,2,*], Arbi Mejri[1], Omrane Kadri[1,3], Florent Kuntz[4], Foued Gharbi[1] and Kais Mannai[5]

[1]Research Unit, Control and Development of Nuclear Technology for Peaceful Uses, National Centre of Nuclear Sciences and Technology, Sidi-Thabet, Tunisia
[2]High Institute of Transport and Logistics, University of Sousse, Tunisia
[3]Department of Radiological Sciences, College of Applied Medical Sciences, King Saud University, Riyadh, Saudia Arabia
[4]Aérial, Centre de Ressources Technologiques, Parc d'Innovation, Illkirch, France
[5]Research Unit, Nuclear and High Energy Physics, Department of Physics, Faculty of Sciences, Tunis, El-Manar University, Tunisia

ABSTRACT

Cobalt-60 gamma irradiation Pilot Plant has been put into operation in 1999 at the National Centre of Nuclear Sciences and Technology (CNSTN), Sidi-Thabet, Tunisia. An initial characterization of this Pilot Plant was performed in order to control technical specification and to determine the overall performance of the irradiator in delivering absorbed dose for sterilization of medical devices and food irradiation. A new irradiation holder was recently installed; it was designed especially for the irradiation of pupae of the Mediterranean fruit fly. It consists of four turn plates which makes it possible to rotate the canisters holding the pupae within the radiation field. The axis of rotation is vertical and parallel to the source pencils. Prior to routine irradiation using the new irradiation holder, validation procedures are necessary to establish conditions of the irradiation within the specification. In the course of these procedures, detailed dose mapping on a vertical plane in the middle of the canister of insect pupae with bulk density of 0.446 g /cm^3 was carried

* E-mail address: kafarah@gmail.com.

out for two irradiation configurations: unturned plates and turned plates. GafChromic dosimeter calibrated against Alanine /ESR dosimetry system was used for the dose measurements. The maximum and minimum dose locations were determined and the dose uniformity ratio calculated and discussed. Detailed analyses of the isodose curves and histogram of the frequency distribution of absorbed dose were also given. Transit dose and dose rate in the reference position inside the canister were measured using Fricke dosimeters. The results of measurements of absorbed dose and dose distribution in insect pupae do not show any significant difference in the dose uniformity ratio $(D.U.R. = \frac{D_{max}}{D_{min}})$ between the two irradiation configurations. At the same time we observed with turned plate's configuration an improvement of the homogeneity of the absorbed dose distribution in the insect pupae showed by the increasing of the pupae irradiated at the minimum dose by about 17 %.

INTRODUCTION

The sterile insect technique is a method of biological control, whereby overwhelming numbers of sterile insects are released. It is a technique which respecting the environment and allows the removal or eradication of pests. Insect pupae, when exposed to ionizing radiation, become sexually sterile because of the introduction of dominant mutations in sperm and ovaries. The sterile males compete with the wild males for female insects. If a female mates with a sterile male then it will have no offspring, thus reducing the next generation's population. Repeated release of insects can eventually wipe out a population, though it is often more useful to consider controlling the population rather than eradicating it. The technique was pioneered in the 1950s in the United States.

The Mediterranean fruit fly (Ceratitis capitata) is the key pest of citrus in the Mediterranean Basin including Tunisia. Tunisia grows 13,000 hectares of citrus with an annual production of around 200,000 tons valued at ca. US $59 million. Despite repeated malathion-bait applications the residual damage is still in the order of 10 to 15%. In addition, Malathion use will be phased out in Tunisia in the near future and the European Union is pressing trading partners to comply with insecticide residue levels. Given this problems and the plan of the government of Tunisia to significantly expand its citrus production areas, mass-rearing and sterilization facility was built and has been in operations since 2003 [1].

This project established with support of the International Atomic Energy Agency (IAEA) and the World Food and Agriculture Organization (FAO). It seeks to control the Mediterranean fruit fly and maintain economically tolerable levels, as a first step, in a pilot area of 6000 hectares. Production of sterile males is in pilot production of sterile males of the Mediterranean fruit fly, 300 m^2 in area, established the CNSTN, Sidi-Thabet. The facility has a maximum rearing capacity of 12 million pupae per week. The selected area was Beni Khalled, in the region of Nabeul in the north of Tunisia, where citrus production is very intensive and where 4 to 5 aerial bait sprays applications are routinely made from September to November immediately followed by sterile male's release. One thousand sterile males per hectare are released weekly in the commercial area and 2000 sterile male per hectare in the buffer area to protect from reinfestations. A control area was also included in the program with 2500 ha of citrus [1].

Accurate dosimetry is needed for dose verification and quality assurance. If the pupae are insufficiently irradiated, they may not be rendered sterile; and, if released, an ecological and economic catastrophe could result. On the other hand, insects over-dosed with radiation, although rendered sterile, are adversely affected in terms of mating competitiveness as compared to non-irradiated flies [2].

A new irradiation holder (4 turned plates) was recently installed; it was designed especially for the irradiation of pupae of the Mediterranean fruit fly. It consists of four turn plates which makes it possible to rotate the canisters holding the pupae within the radiation field. The axis of rotation is vertical and parallel to the source pencils. In order to optimize the positioning of the new insect irradiation holder inside the irradiation cell by obtaining the best homogenize absorbed dose, dose rate measurements in air were carried out: Determination of the position of the source pencils in the irradiation position by measurement of the dose distribution along the protective cylinder, Control of the isotropy of gamma radiation emission, Dose distribution in air in vertical plane parallel to the source pencils relative to the positions of the product carriers. The simulation data has been also obtained using the GEANT4 Monte Carlo code of CERN [3]. This code has been validated, inside the Tunisian facility, as efficient tool of dose rate measurement prediction inside gamma industrial irradiation facilities [4] and it was used for the optimization of some gamma irradiation processing parameters [5-9].

In this chapter, detailed dose mapping in air and inside the canister of insect pupae with bulk density of 0.446 g /cm^3 was carried out for two irradiation configurations: unturned plates and turned plates. The maximum and minimum dose locations were determined and the dose uniformity ratio calculated and discussed. Detailed analyses of the isodose curves and histogram of the frequency distribution of absorbed dose were also given.

2. Materials and Methods

2.1. Discription of the Pilot-scale Co-60 Gamma-Irradiation Plant

The Tunisian pilot-scale Co-60 gamma-irradiation facility is designed for conservation of foodstuff and sterilisation of medical devices. The source consists of eight encapsulated cobalt-60 source pencils with a diameter of 9.7 mm. The overall length is 452 mm. The initial activity of the source is 98.000 Ci (9/4/99). To irradiate products in industrial configuration, the installation is equipped with a stainless steel telescopic source rack. This source rack makes it possible to obtain a linear source of approximately 900 mm height. The source pencils are distributed on a diameter of 140 mm for the higher source rack and of 80 mm for lower one. The source rack comprises 20 housing allowing a loading and a use of the sources during several years. These sources are stored in dry conditions in a cylindrical container in which they were transported. It consists of steel and lead. The irradiation facility consists of a concrete shielding including the irradiation cell with a protective labyrinth, a conveyor system, a control room, a dosimetry laboratory, a warehouse for ionised and non-ionised products and refrigerated rooms. The products to be treated are transported inside the irradiation cell using 5 carriers moved by electromechanical convoying system fixed on the ground [10] (Figure 1).

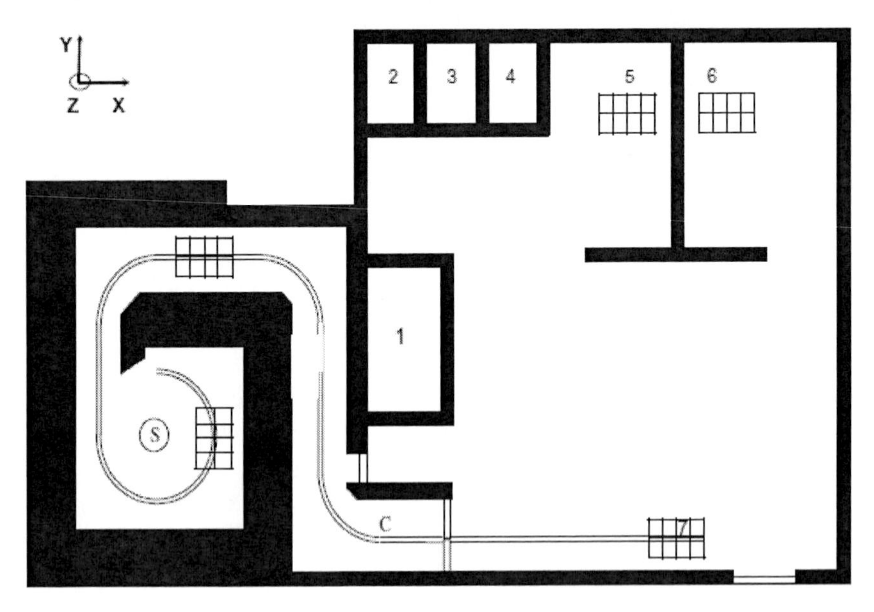

Figure 1. Schematic representation of the Tunisian gamma irradiation facility: (S) source rack; (C) conveyor; 1, control room; 2, EPR laboratory; 3, dosimetry laboratory; 4, maintenance group; 5, inlet storage; 6, outlet storage; 7, carrier and walls.

2.2. The New Irradiation Holder

It consists of four turn plates with 19 cm in diameter; they are installed on a circle of radius 42 cm around the source which makes it possible to rotate the canisters holding the pupae within the radiation field. The number of revolution is 9 turns per minute. The axis of rotation is parallel to the source rack. The turn plates were elevated at the height of 144 cm (Figure 2).

2.3. The Irradiation Canister

It is an aluminum cylinder with a diameter of 13 cm and a height of 24 cm designed to the insect pupae irradiation purposes.

2.4. Dosimetry Systems

Dosimetry systems were used for the dose distribution measurements in air are Red and Amber Perspex, Cellulose Triacetate (CTA), Alanine/EPR dosimeter and Fricke dosimeter. Red (type 4034, batch FB and FC , dose range : 5-50 kGy) and Amber (3042 type, batch M, dose range : 1-15 kGy) Perspex dosimeters, sealed in polyethylene-aluminium sachets (produced by Harwell, U.K.) [11], in plant-calibrated against Alanine/EPR dosimetry system [12,13], were evaluated at 640 and 603 nm, respectively, using a Aerial Optical Dosimetry System [14]. CTA (type FTR-125, dose range: 0.1-100 kGy) film is manufactured for dosimetry by Fuji Photo Film Co [15]. Its nominal thickness is 0.125 mm and its width is 8 mm. The CTA films and Fricke solutions [16] are respectively read at 280 nm and 305 nm using the same Aerial Optical Dosimetry System.

GafChromic dosimeter was used for the dose distribution measurements in the insect pupae (dose range 1-5 kGy [17] in-plant calibrated against Transfer Alanine /EPR dosimetry system. Dosimeters were evaluated at 580 nm wavelength using Aerial Optical Dosimetry System. GafChromic dosimeters were placed at a defined network (Figure 3).

2.5. Gridding Method

The gridding method selected was the software method: SURFER 7.0. This method is geostatistical gridding method, which produces visually contours and surface plots from irregularly spaced data. Kriging can be custom-fit to data set by specifying the appropriate variogram model. The mathematical variogram model specifies the spatial variability of the data set and resulting grid. The interpolation weights, which are applied to data points during the grid node calculations, are direct functions of the variogram model. The linear variogram model was chosen with a scale of 1.

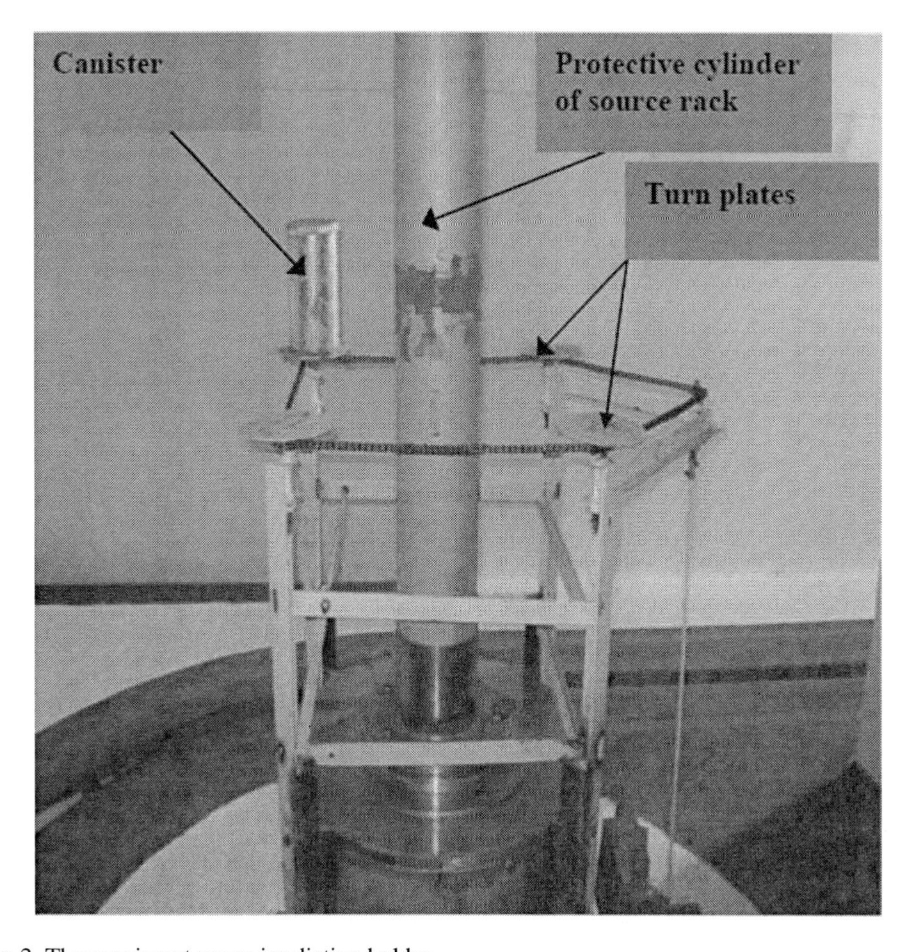

Figure 2. The new insect pupae irradiation holder.

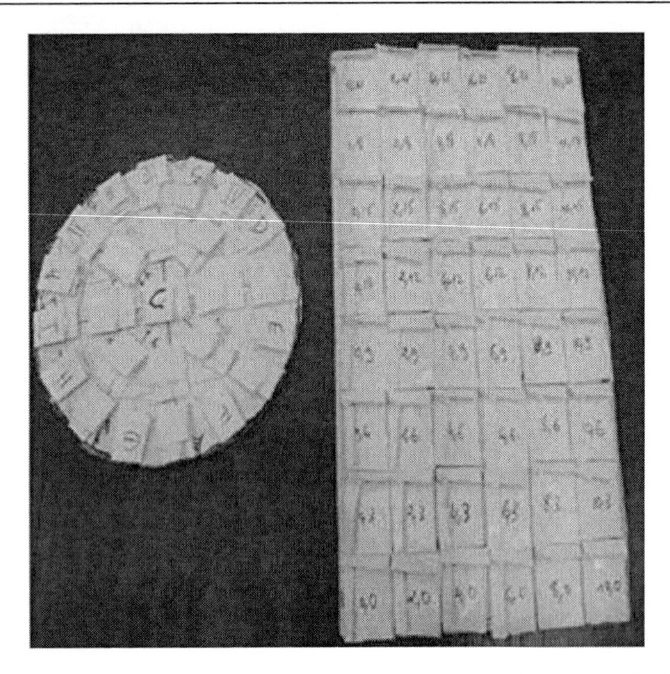

Figure 3. Localization of GafChromic dosimeters in the vertical (right figure) and the horizontal (left figure) plans.

2.6. GEANT 4 Simulation Code

GEANT is a Detector Description and Simulation Tool system developed by the Application Software Group at CERN. The main applications of the GEANT code are the transport of particles through an experimental set-up for a simulation of detector response and particle trajectories. The code is written in user-friendly way with graphics support that enables on-line monitoring of the geometrical details of the detector system. GEANT code, initially designed for high-energy physics has proven to be very efficient for low-energy physics. The code is mainly a library of routines written in Fortran77 language, offered to the user for the construction of the specific applications.

3. Results and Discussion

3.1. Dose Distribution in Air

3.1.1. Determination of the Source Pencils Position

The determination of the position of source pencils in the irradiation position by dose distribution measurements of along the protective cylinder is very important to optimise the positioning of the insect irradiation holder (4 turned plates) in the irradiation cell in order to homogenise the absorbed dose. In irradiation position, the source length was deployed to 88.5 cm. We placed against the protective cylinder (Figure 4) CTA film strips of 177 cm long, packed in a polyethylene sachet sealed by heat. The irradiation time was two hours. The results of measurements show that the height of the middle of the source (corresponding to the maximum of the dose rate) is at about 155 cm from the ground of irradiation cell (Figure 4). We can also affirm that the source is located between 111 and 199 cm.

3.1.2. The Vertical Dose Rate Variation

Figure 5 shows a comparison between measured and calculated dose rate along the vertical direction from the ground to the upper limit of the irradiation cell. As seen in the figure, dose rate increases as the measurement points are close to the middle of the source at 155 cm high, which is well reproduced by simulation. Although, it is interesting to note that for some few points difference between measurement and calculation results exceeds 6%.

This is related to the errors in experimental dosimeters placement, which was not taken into account in the simulation [4].

3.1.3. The Horizontal Dose Rate Variation

For this experiment, Amber dosimeters are fixed on a horizontal axis perpendicular to the source pencils at the height of 155 cm at the following distances: 20, 25, 30, 40.....260 cm from the source axis. Three dosimeters were irradiated at each measurement point. Irradiation time is 90 minutes for the dosimeters placed between 20 and 80 cm and 750 minutes for those placed between 90 and 260 cm.

This experiment shows a very significant decrease of the dose rate with the distance from the source. Indeed, while moving away from 20 to 60 cm of the source, the dose rate decreases 83 % whereas between 220 and 260 cm from the source the loss of dose rate is about 13 % (Figure 6). This aspect is translated directly into term of quality of processing (homogeneity). The nearer the product is to the source, the less homogeneity of the absorbed dose in the product. This disadvantage is accompanied nevertheless by the advantage of very high dose rate.

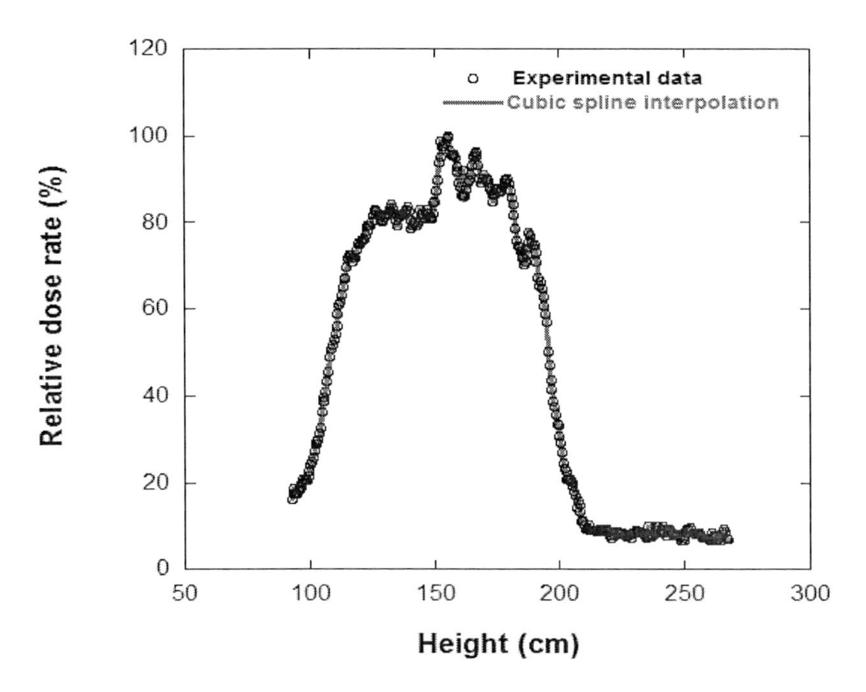

Figure 4. Dose rate distribution along the protective cylinder.

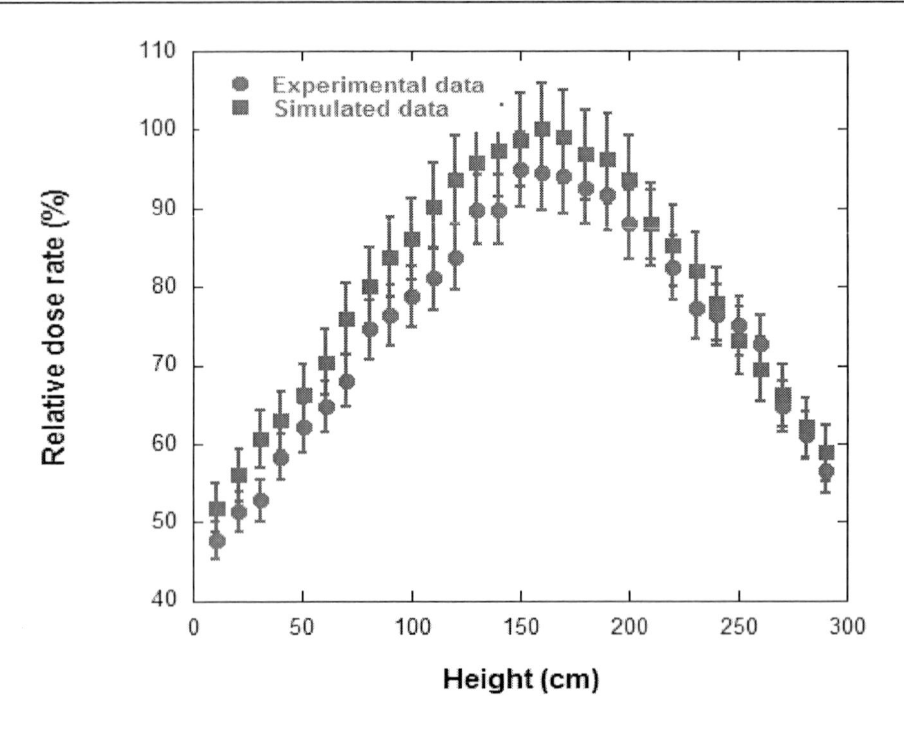

Figure 5. Experimental and calculated dose rate distribution in air along a vertical axis parallel to the source pencils.

3.1.4. Isotropy Control of Radiation Emission

The first experiment consists of measuring the dose rate in the irradiation room around the source on the horizontal plan at 155 cm height. In this experiment, Amber dosimeters were placed on eight horizontal axis perpendiculars to the source pencils with the following angles: 0, 45, 90, 135, 180, 225 and 270 degree and at the distances of 40, 125, 170 and 220 cm from the source axis. Three dosimeters were irradiated at each measurement point. The irradiation time is 60 minutes for the dosimeters placed at 40 cm and 720 minutes for those placed between 125 and 220 cm. The maximum anisotropy does not exceed 6 % between 0 and 360 degree around the source. Considering that the uncertainty of the measurements is 6 %, we can assume that the source emits an isotropic gamma radiation. The experimental dose values corresponding to an 8 x 4 mapping were used to determine a new interpolated grid with added theoretical dose values closely fitting the experimental data. Upon completion of this optimisation procedure we finally computed 8 x 11 grid of dose values to built the isodose curve using a geostatistical griding method, the kriging method. Figure 7 shows dose rate distribution obtained in this plane for experimental (Figure (a)) and for simulation (Figure (b)) results. The decrease of dose rate as function of the distance to the source is reproduced by simulation with good agreement [4, 10].

The second experiment consists of measuring the dose rate around the source. Alanine and Fricke dosimeters were placed together in the centre of each turn plate and then irradiated. The results of the dosimetric measurements are summarized in Table 1. The maximum anisotropy does not exceed 4 % on the four turn plates around the source, which enables us to conclude that the source emission is isotropic.

Table 1. Anisotropy evaluation of gamma radiation emission around the source on the turn plates

	Average dose rate ($Gy.S^{-1}$)				Standard Deviation (%)
Trun Table N°	1	2	3	4	
Fricke	0.615	0.595	0.639	0.628	3.52
Alanine	0.618	0.642	0.634	0.626	1.62

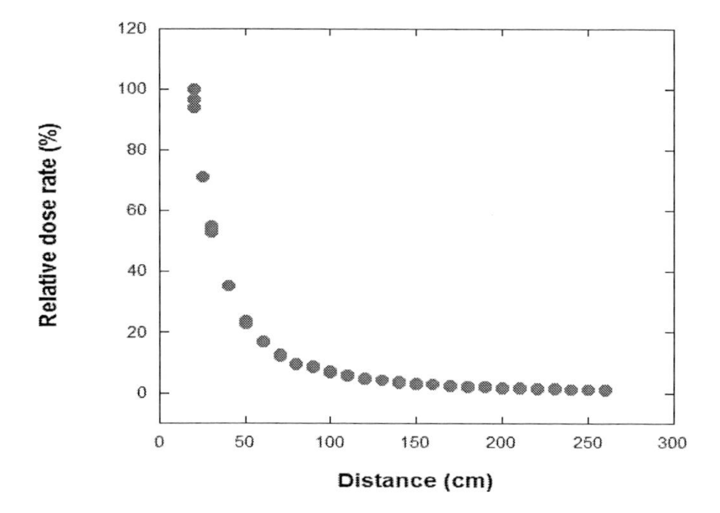

Figure 6. Dose rate evaluation in air in a horizontal plane perpendicular to the source pencils with distance.

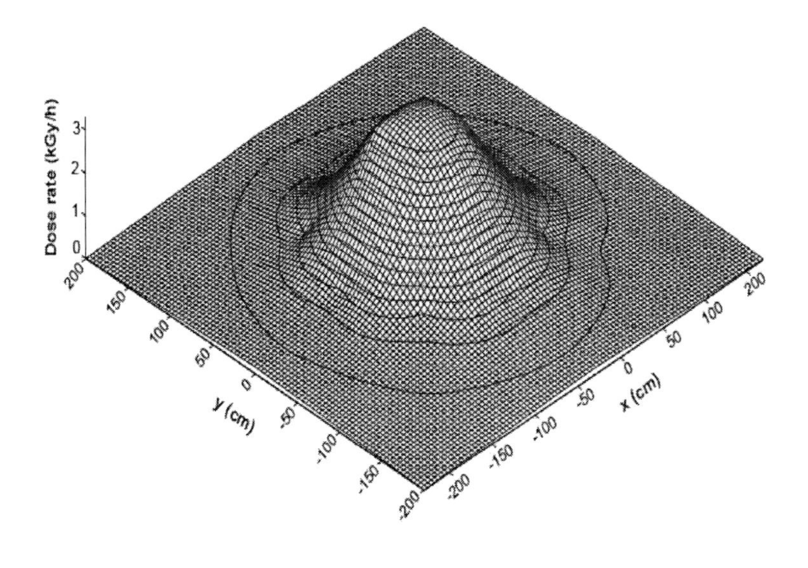

(a)

Figure 7. (Continued).

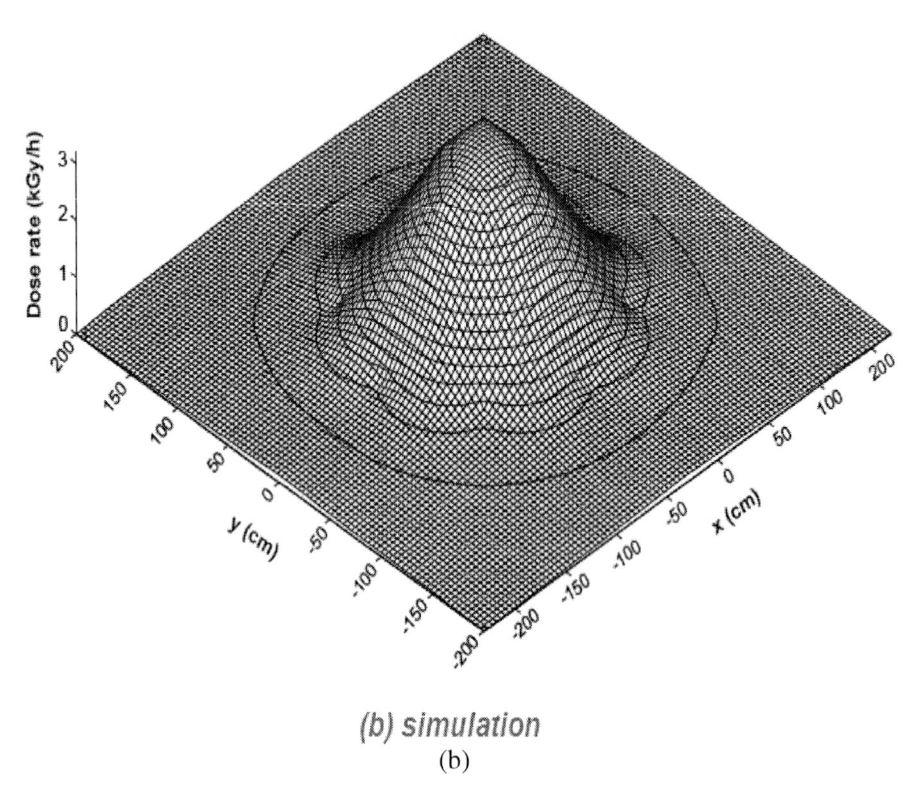

(b)

Figure 7. Three-dimensional view of dose rate distribution in a horizontal plane perpendicular to the source pencils at 155 cm of high from the ground respectively, for experimental (a) and simulation results (b).

3.2. Dose Distribution in the Insect Pupae

The aim of these experiments is the determination of similar dose zones and dose extrema zones (minimum ($D_{\mathbf{min}}$) and maximum ($D_{\mathbf{max}}$) absorbed doses) and dose uniformity ratio $D.U.R. = \frac{D_{\mathbf{max}}}{D_{\mathbf{min}}}$ for the two irradiation processes (unturned and turned plates).

Zones that are statistically equivalent to the minimum zone dose are found by comparing the mean dose for the zone, $\bar{D}_z$, with the value: $\bar{D}_{\mathbf{min}} + \delta$ [18], where δ is the minimum detectable difference.

The difference between the dose means must be greater than the minimum detectable difference or least significant difference to be statistically significant. Zones with $\bar{D}_z$ less than this value are statistically equivalent to the minimum dose zone. Zones that are statistically equivalent to the maximum zone dose are found by comparing the mean dose for the zone, $\bar{D}_z$, with the value : $\bar{D}_{\mathbf{max}} - \delta$. Zones with $\bar{D}_z$ greater than this value are statistically equivalent to the maximum dose zone.

3.2.1. Unturned Plates Irradiation Process

Figures 8 and 9 represent respectively the isodose curves in the horizontal plan perpendicular to source rack in the middle of the cylindrical canister and the isodose curves in vertical plan parallel to the source rack across of the cylindrical canister. As we can seen in these figures, the maximum absorbed dose values are located in the middle of the cylindrical canister and the absorbed dose gradients across the canister decreases to the peripheral regions near the source pencils.

Figure 10 shows the isodose curves in vertical plan parallel to the source rack at the centre of the cylindrical canister. The majority of the dose values in central regions of the cylinder are close to the minimum dose, varying from the target dose by no more than 9 %.

The dose values in the entire cylinder vary from the target minimum dose by about 57 %.

3.2.2. Turned Plates Irradiation Process

Figure 11 represents the isodose curves in a vertical plan placed in the middle of the cylindrical canister. The majority of the dose values in the entire cylinder are close to minimum dose, varying from the target dose by no more than 11 %.

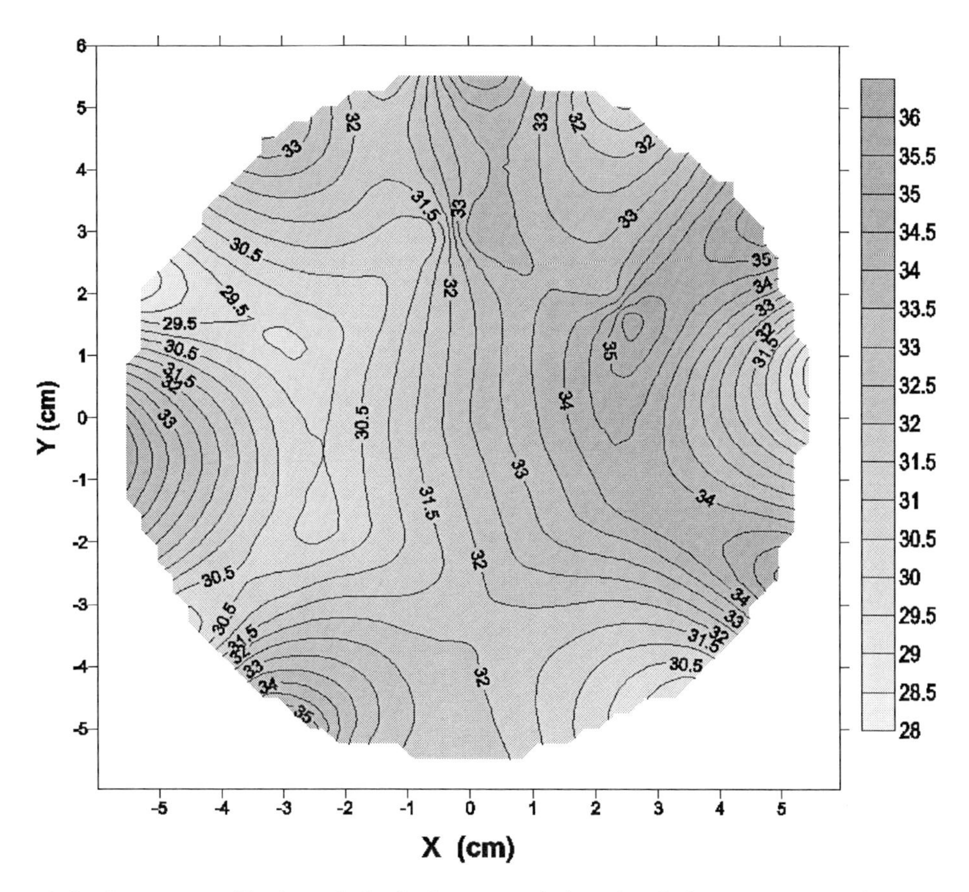

Figure 8. Isodose curves of horizontal plan in the unturned plates irradiation process (Gy/min).

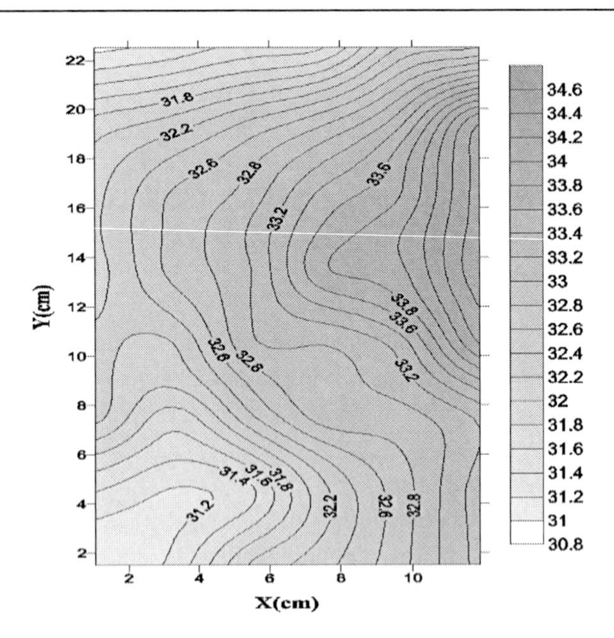

Figure 9. Isodose curves of vertical plan across of the cylindrical canister in the unturned plates irradiation process (Gy/min).

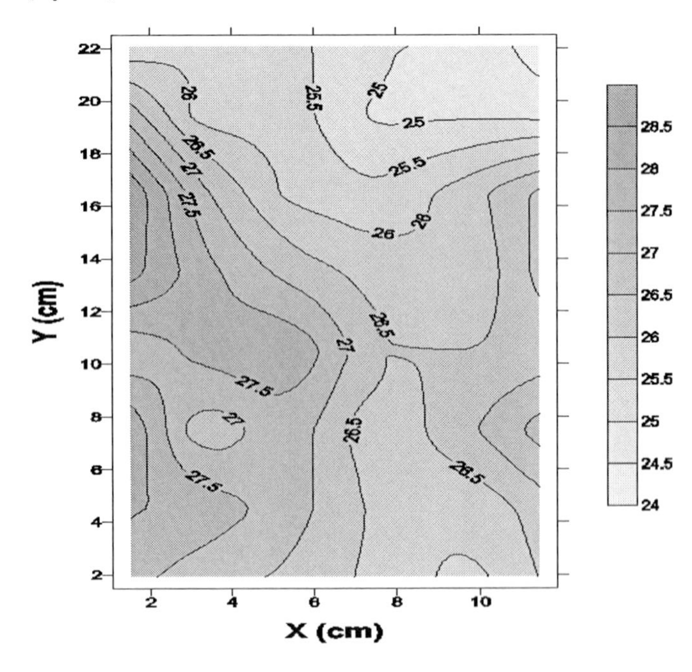

Figure 10. Isodose curves of vertical plan in the unturned plates irradiation process (Gy/min).

The values of the dose uniformity ratio (D. U. R.) were calculated for the unturned and the turned plate irradiation processes are respectively 1.20 and 1.07.This is about 10 % lower than the value of (D. U. R.) for the unturned plates irradiation process, which represents a significant improvement of the homogeneity of the absorbed dose distribution in the insect pupae.

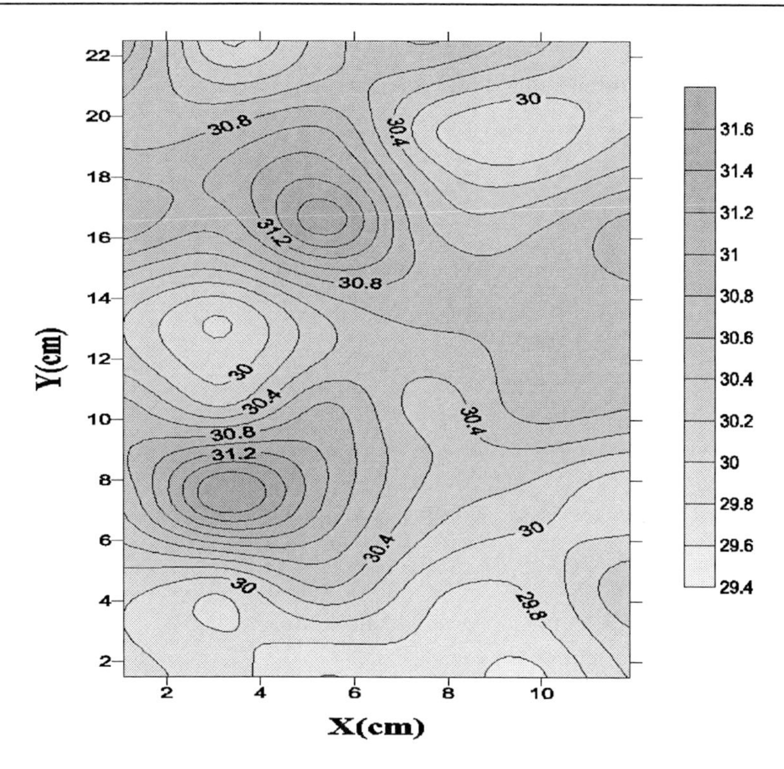

Figure 11. Isodose curves of vertical plan in the turned plates irradiation process (Gy/min).

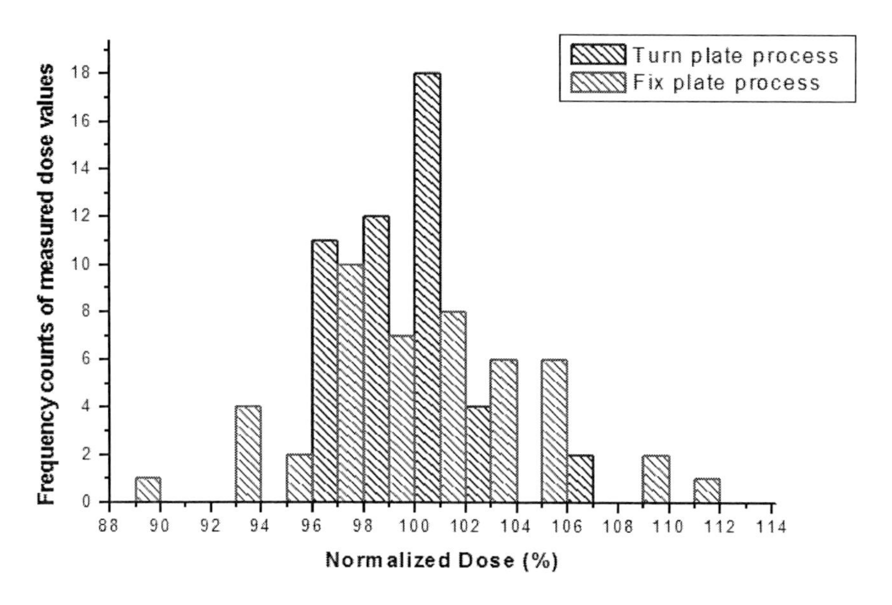

Figure 12. Frequency of observed normalized doses values for the two irradiation processes.

Histogram representation of the frequency distribution of dose measured for the two irradiation processes can be seen in Figure 12. For the unturned plate irradiation process, only 15% of the dose values are close to the target minimum dose values. For the turned plate irradiation process, 49% of the dose values are close to the target minimum dose values.

The frequencies counts of maximum dose values measured are respectively 6.4% and 4.25% for the unturned and the turned plate processes.

Conclusion

The absorbed dose distribution measurements in air carried out in order to optimize the positioning of the new irradiation holder around the source rack are in good agreement with calculations made by Geant 4 code.

The isodose curves provide detailed spatial information on the dose rate levels, and give a general indication of dose rate levels throughout the canister volume for the two irradiation processes of insect pupae with the new irradiation holder installed at the Tunisian Cobalt-60 gamma irradiation pilot-scale facility. The results of measurements of absorbed dose and dose distribution in insect pupae show an improvement in the dose uniformity ratio (D.U.R.) between the two irradiation processes by about 10%. At the same time we observed for the unturned plate irradiation process, only 15% of the dose values are close to the target minimum dose values. While, for the turned plate irradiation process, 49% of the dose values are close to the target minimum dose values.

References

[1] *CNSTN.* http://www.cnstn.rnrt.tn/male_sterile_ceratite_projets.php.

[2] M. L. Walker, W. L. McLaughlin, J. M. Puhl, P. Gomez. *Appl. Radiat. lsot.* 48, 117 (1997).

[3] S. Agostinelli et al. *Nucl. Instr. Meth.* A. 506, 250 (2003).

[4] F. Gharbi, O. Kadri, K. Farah, K. Mannai, *Rad. Phys. Chem.* 74, 102 (2005).

[5] O. Kadri, F. Gharbi, K. Farah, *Nucl. Instr. Meth.* B. 239, 391 (2005).

[6] O. Kadri, F. Gharbi, K. Farah, K. Mannai, A. Trabelsi. *App. Radiat. Isot.* 64, 170 (2006).

[7] O. Kadri, F. Gharbi, A. Trabelsi. *Nucl. Instr. Meth.* B. 245, 459 (2006).

[8] F. Gharbi, O. Kadri, A. Trabelsi. *Nucl. Instr. Meth.* A. 566, 516 (2006).

[9] K. Mannai, B. Askri, A. Loussaief, A. Trabelsi, 2007. *App. Radiat. Isot.* 65, 701 (2007).

[10] K. Farah, T. Jerbi, F. Kuntz, A. Kovacs, 2006. *Rad. Meas.* 41, 201 (2006).

[11] ISO/ASTM, *Practice for Use of a Polymethylmethacrylate Dosimetry System*, ISO/ASTM Standard 51276, American Society for Testing and Materials, Philadelphia, PA. (2004).

[12] ISO/ASTM, *Practice for Use the Alanine-EPR Dosimetry Systems*, ISO/ASTM Standard 51607, American Society for Testing and Materials, Philadelphia, PA. (2002).

[13] ISO/ASTM, Standard Guide for Selection and Calibration of Dosimetry Systems for Radiation Processing, ISO/ASTM Standard 51261. American Society for Testing and Materials, Philadelphia, PA (2004).

[14] *Arial*, http://www.aerial-crt.com/en/indus-dosimetry-equipment-supply-/aerode.html.

[15] ISO/ASTM, *Practice for Use of Cellulose Acetate Dosimetry Systems*, ISO/ASTM Standard 51650, American Society for Testing and Materials, Philadelphia, PA. (2004).

[16] ASTM, *Practice for Using the Fricke Reference Standard Dosimetry System* (ASTM Standard E1026, American Society for Testing and Materials, Philadelphia, PA. (2004).

[17] ISO/ASTM, *Standard practice for Use of Radiochromic Film Dosimetry System*, ISO/ASTM Standard 51275, American Society for Testing and Materials, Philadelphia, PA.(2004).

[18] ASTM. *Standards guide for absorbed dose mapping in radiation processing facilities*. ASTM Standard WK348-2003, American Society for Testing and Materials, Philadelphia, Pa. (2003).

INDEX

A

B

C

D

E

F

G

H

I

N

O

P

Q

R

S

T

U

V

W

X

Y

Z